Inorganic Nanomaterials from Nanotubes to Fullerene-like Nanoparticles

Fundamentals and Applications

World Scientific Series in Nanoscience and Nanotechnology

Series Editor-in-Chief: Frans Spaepen (*Harvard University, USA*)

Vol. 1 Molecular Electronics: An Introduction to Theory and Experiment
Juan Carlos Cuevas (*Universidad Autónoma de Madrid, Spain*) *and*
Elke Scheer (*Universität Konstanz, Germany*)

Vol. 2 Nanostructures and Nanomaterials: Synthesis, Properties, and Applications, 2nd Edition
Guozhong Cao (*University of Washington, USA*) *and*
Ying Wang (*Louisiana State University, USA*)

Vol. 3 Molecular Cluster Magnets
edited by *Richard Winpenny* (*The University of Manchester, UK*)

Vol. 4 Plasmonics and Plasmonic Metamaterials: Analysis and Applications
edited by *Gennady Shvets* (*The University of Texas, Austin, USA*)
and Igor Tsukerman (*The University of Akron, USA*)

Vol. 5 Inorganic Nanomaterials from Nanotubes to Fullerene-Like Nanoparticles: Fundamentals and Applications
Reshef Tenne (*Weizmann Institute of Science, Israel*)

Volume 5

World Scientific Series in
Nanoscience and Nanotechnology

Image courtesy of Dr. Maya Bar-Sadan

Inorganic Nanomaterials from Nanotubes to Fullerene-like Nanoparticles

Fundamentals and Applications

Reshef Tenne
Weizmann Institute of Science, Israel

NEW JERSEY · LONDON · SINGAPORE · BEIJING · SHANGHAI · HONG KONG · TAIPEI · CHENNAI

Published by

World Scientific Publishing Co. Pte. Ltd.
5 Toh Tuck Link, Singapore 596224
USA office: 27 Warren Street, Suite 401-402, Hackensack, NJ 07601
UK office: 57 Shelton Street, Covent Garden, London WC2H 9HE

British Library Cataloguing-in-Publication Data
A catalogue record for this book is available from the British Library.

World Scientific Series in Nanoscience and Nanotechnology — Vol. 5
INORGANIC NANOMATERIALS FROM NANOTUBES TO FULLERENE-LIKE NANOPARTICLES
Fundamentals and Applications

Copyright © 2013 by World Scientific Publishing Co. Pte. Ltd.

All rights reserved. This book, or parts thereof, may not be reproduced in any form or by any means, electronic or mechanical, including photocopying, recording or any information storage and retrieval system now known or to be invented, without written permission from the Publisher.

For photocopying of material in this volume, please pay a copying fee through the Copyright Clearance Center, Inc., 222 Rosewood Drive, Danvers, MA 01923, USA. In this case permission to photocopy is not required from the publisher.

ISBN 978-981-4343-38-1

Printed in Singapore by Mainland Press Pte Ltd.

GENERAL INTRODUCTION

The present book is a judicious assembly of some 45 of the most important publications of Prof. R. Tenne from his work over the last two decades. These publications are focused on the study of closed-cage and hollow nanostructures from inorganic compounds with layered (2-D) structure. These nanostructures come generally as multiwall quasi-spherical nanoparticles- the so-called inorganic fullerene-like (IF) or as inorganic nanotubes (INT). As such they may be considered as the extension of the well-known carbon fullerenes and carbon nanotubes. Before going into a detailed discussion of the virtues of this new class of materials, a short background into the discipline of hollow nanostructures is desirable.

The chemical bond is unstable beyond a distance of 2–2.5 Å and even weak chemical forces, such as the van der Waals or hydrophobic interactions become insignificant beyond 5–7 Å. Generally speaking, therefore, chemistry is not favorable to open spaces and hence most crystalline materials are compact and do not contain voids or hollow space. Nonetheless, already in the beginning of the last century chemists realized that some chemical moieties adopt hollow closed-cage structure. Thus, borohydride ions of the form $B_nH_n^{2-}$ and carbaboranes, i.e. molecules consisting of boron-carbon and hydrogen atoms, were among the first studied examples of polyhedral structures with an empty core [see Ref. 1]. The stability of these polyhedral structures was attributed to the three atom B-H-B bond, which permits the electron deficient boron atoms to form spatially stable polyhedra with hollow core.

Also early on, it was found that asbestos minerals, like chrysotile (which in its flat form is named lizardite), halloysite, kaolinite, etc. accommodate tubular structures with hollow core. For example, the tendency of kaolinite sheet to fold into tubular structures was studied by Pauling [2]. The stimulus for the curving was attributed to the built-in asymmetry of the layered structure along the c-axis. Each molecular layer of chrysotile (kaolinite) consists of a fused sheet of silica tetrahedra and a sheet of magnesia (alumina) octahedra which are fused together via a common oxygen atom (see **Fig. 1a**). The overall formula of chrysotile is $Mg_3Si_2O_5(OH)_4$ and is $Al_2Si_2O_5(OH)_4$ for kaolinite. The size difference in the (a,b) plan between the two interconnected layers stipulates that the silica tetrahedra occupy the outer face of the scrolling sheet and they are under tensile stress. On the other hand the inner concave alumina octahedra of the asbestos sheet are subdued to compressive stress. This asymmetry leads to folding and scrolling with a clear cut energy minimum as a function of the nanotube (nanoscroll) radius [3]. **Fig. 1b** shows a schematic

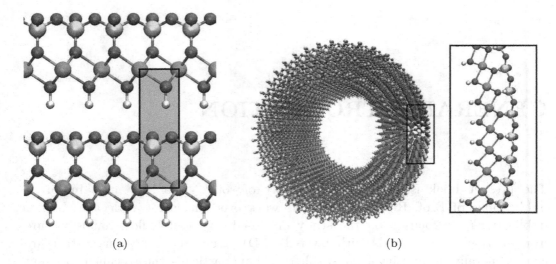

Fig. 1. (a) Schematic side-view (010) of the mineral lizardite (chrysotile in the tubular form); (b) Schematic drawing of a 3-wall chrysotile nanotube (courtesy of Dr. A.N. Enyashin).

Fig. 2. High-resolution TEM image of the tip of a chrysotile nanotube (courtesy of Prof. N. Roveri and Dr. R. Popovitz-Biro).

rendering of multiwall chrysotile nanotube with clear hollow core in its center, while **Fig. 2** shows a transmission electron microscopy (TEM) of such nanotube. Much attention has been focused on the toxicity of these threaded nanostructures. Interestingly enough though, synthetically produced chrysotile ($Mg_3Si_2O_5(OH)_4$) nanotubes were found to be non-toxic in recent studies [4]. This observation suggests that the asbestos toxicity is rooted in its specific chemical interactions with the human tissues or certain impurities, like iron, and cannot be attributed entirely to its small diameter and large aspect ratio (length to diameter ratio), alone.

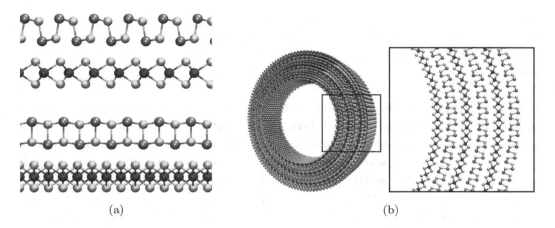

Fig. 3. (a) Schematic drawing of a side-view (010) structure of the "misfit" compound PbS/NbS$_2$ in the planar and (b) in the tubular form (courtesy of Dr. A.N. Enyashin).

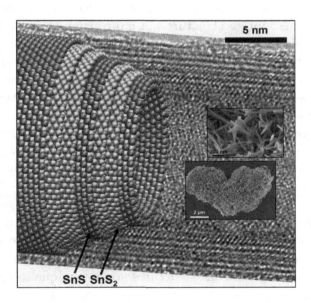

Fig. 4. A high resolution TEM picture of a superstructure SnS/SnS$_2$ ("misfit") nanotube. Superimposed on the TEM image is a drawing of the different layers of the nanotube (Sn-grey; S-yellow). In the inset are two SEM images of a pile of such nanotubes in two magnifications (courtesy of Mr. G. Radovsky and Prof. G. Seifert).

Another series of synthetic layered compounds with a "built-in" asymmetry, the so-called "misfit" compounds, are known to form tubular structures with hollow core. Here monomolecular sheets of compounds with cubic lattice, like PbS, are intercalated between the layers of, e.g., NbS$_2$ [5]. **Fig. 3a** shows a schematic drawing of a PbS/NbS$_2$ misfit compound. **Fig. 3b** shows an artist rendering of a multiwall nanotube of the same misfit compound. **Fig. 4** shows a TEM image of the misfit SnS/SnS$_2$ nanotube and schematic drawing of this nanotube overlayed on the TEM picture. This artificial superstructure is characterized by a mismatch between the a-b lattices of the two compounds (SnS and SnS$_2$). In analogy to asbestos crystals,

[15] J.A. Wilson and A.D. Yoffe, Transition metal dichalcogenides discussion and interpretation of observed optical, electrical and structural properties, *Adv. Phys.* **269**, 193–335 (1969).

[16] F. Lévy, *Structural Chemistry of Layer-Type Phases*, Vol. 5, D. Reidel, Dordrecht (1976).

[17] R. Tenne, L. Margulis, M. Genut, and G. Hodes, Polyhedral and cylindrical structures of tungsten disulphide, *Nature* **360**, 444–446 (1992).

[18] L. Margulis, G. Salitra, R. Tenne, and M. Talianker, Nested fullerene-like structures, *Nature* **365**, 113–114 (1993).

[19] R. Tenne, Doped and heteroatom-containing fullerene-like structures and nanotubes, *Adv. Mater.* **7**, 965–995 (1995).

[20] L. Margulis, S. Iijima, and R. Tenne, Nucleation of WS_2 fullerenes at room temperature, *Microscopy, Microanalysis, Microstructures* **7**, 87–89 (1996).

[21] P.A. Parilla, A.C. Dillon, K.M. Jones, G. Riker, D.L. Schulz, D.S. Ginley, and M.J. Heben, The first true inorganic fullerenes? *Nature* **397**, 114 (1999).

[22] P.A. Parilla, A.C. Dillon, B.A. Parkinson, K.M. Jones, J. Alleman, G. Riker, D.S. Ginley, and M.J. Heben, Formation of nanooctahedra in molybdenum disulfide and molybdenum diselenide using pulsed laser vaporization, *J. Phys. Chem B* **108**, 6197–6207 (2004).

[23] R.R. Chianelli, E.B. Prestridge, T.A. Pecoraro, and J.P. Deneufville, Molybdenum disulfide in the poorly crystalline "rag" structure, *Science* **203**, 1105–1107 (1979).

[24] J.V. Sanders, Structure of catalytic particles, *Ultramicroscopy* **20**, 33–37 (1986).

[25] a. M. Nath and C.N.R. Rao, New metal disulfide nanotubes, *J. Am. Chem. Soc.* **123**, 4841–4842 (2001); b. P. Li, C.L. Stender, E. Ringe, L.D. Marks, and T.W. Odom, Synthesis of TaS_2 nanotubes from Ta_2O_5 nanotube templates, *Small* **6**, 1096–1099 (2010).

[26] a. S.Y. Hong, R. Popovitz-Biro, Y. Prior, and R. Tenne, Synthesis of SnS_2/SnS fullerene-like nanoparticles: a superlattice with polyhedral shape, *J. Am. Chem. Soc.* **125**, 10470–10474 (2003); b. A. Yella, E. Mugnaioli, M. Panthöfer, H.A. Therese, U. Kolb, and W. Tremel, Bismuth-catalyzed growth of SnS_2 nanotubes and their stability, *Angew. Chem. Int. Ed.* **48**, 6426–6430 (2009); c. G. Radovsky, R. Popovitz-Biro, M. Staiger, K. Gartsman, C. Thomsen, T. Lorenz, G. Seifert, and R. Tenne, Synthesis of copious amounts of SnS_2 and SnS_2/SnS nanotubes with ordered superstructures, *Angew. Chem. Int. Ed.* **51**, 12316–12320 (2011).

[27] U.K. Gautam, S.R.C. Vivekchand, A. Govindaraj, G.U. Kulkarni, N.R. Selvi, and C.N.R. Rao, Generation of onions and nanotubes of GaS and GaSe through laser and thermally induced exfoliation, *J. Am. Chem. Soc.* **127**, 3658–3659 (2005).

[28] Y. Rosenfeld Hacohen, E. Grunbaum, R. Tenne, J. Sloan, and J.L. Hutchison, Cage structures and nanotubes of $NiCl_2$, *Nature* **395**, 336–337 (1998).

[29] J. Cumings and A. Zettl, Mass-production of boron nitride double-wall nanotubes and nanococoons, *Chem. Phys. Lett.* **316**, 211–216 (2000).

[30] D. Golberg, Y. Bando, C. Tang, and C. Zhi, Boron nitride nanotubes, *Adv. Mater.* **19**, 2413–2432 (2007).

[31] F. Krumeich, H.J. Muhr, M. Niederberger, F. Bieri, B. Schnyder, and R. Nesper, Morphology and topochemical reactions of novel vanadium oxide nanotubes, *J. Am. Chem. Soc.* **121**, 8324–8331 (1999).

[32] T. Kasuga, M. Hiramatsu, A. Hoson, T. Sekino, and K. Niihara, Formation of titanium oxide nanotube, *Langmuir* **14**, 3160–3163 (1998).

[33] G.H. Du, Q. Chen, R.C. Che, Z.Y. Yuan, and L.M. Peng, Preparation and structure analysis of titanium oxide nanotubes, *Appl. Phys. Lett.* **79**, 3702–3705 (2001).

[34] J. Goldberger et al., Single-crystal gallium nitride nanotubes, *Nature* **422**, 599–602 (2003).

[35] Q. Wu, Z. Hu, X. Wang, Y. Lu, X. Chen, H. Xu, and Y. Chen, Synthesis and characterization of faceted hexagonal aluminum nitride nanotubes, *J. Am. Chem. Soc.* **125**, 10176–10177 (2003).

[36] L.W. Yin, Y. Bando, J.H. Zhan, M.S. Li, and D. Golberg, Self-assembled highly faceted wurtzite-type ZnS single-crystalline nanotubes with hexagonal cross-sections, *Adv. Mater.* **17**, 1972–1977 (2005).

[37] W. Stöber, A. Fink, and E. Bohn, Controlled growth of monodisperse silica spheres in micron size range, *J. Colloid Interf. Sci.* **26**, 62–69 (1968).

[38] M. Nakamura and Y. Matsui, Silica gel nanotubes obtained by the sol-gel method, *J. Am. Chem. Soc.* **117**, 2651–2652 (1995).

[39] B.C. Satishkumar et al., Oxide nanotubes prepared using carbon nanotubes as templates, *J. Mater. Res.* **12**, 604–606 (1997).

[40] T. Kasuga, M. Hiramatsu, A. Hoson, T. Sekino, and K. Niihara, Formation of titanium oxide nanotube, *Langmuir* **14**, 3160–3163 (1998).

[41] R. Beranek, H. Hildebrand, and P. Schmuki, Self-organized porous titanium oxide prepared in H_2SO_4/HF electrolytes, *Electrochem. Solid-State Lett.* **6**, B12–B14 (2003).

[42] D. Gong, C.A. Grimes, O.K. Varghese, W. Hu, R.S. Singh, Z. Chen, and E.C. Dickey, Titanium oxide nanotube arrays prepared by anodic oxidation, *J. Mater. Res.* **16**, 3331–3334 (2001).

[43] K. Zhu, T.B. Vinzant, N.R. Neale, and A.J. Frank, Removing structural disorder from oriented TiO_2 nanotube arrays: reducing the dimensionality of transport and recombination in dye-sensitized solar cells, *Nano Lett.* **7**, 3739–3746 (2007).

[44] a. R. Saito, M. Fujita, G. Dresselhaus, and M.S. Dresselhaus, Electronic structure of graphene tubules based on C_{60}, *Phys. Rev. B* **46**, 1804–11 (1992); b. J.W. Mintmire, B.I. Dunlap, and C.T. White, Are fullerene tubules metallic? *Phys. Rev. Lett.* **68**, 631–634 (1992); c. N. Hamada, S. Sawada, and A. Oshiyama, New one-dimensional conductors: graphitic microtubules, *Phys. Rev. Lett.* **68**, 1579–1581 (1992).

[45] a. G. Seifert, H. Terrones, M. Terrones, G. Jungnickel, and T. Frauenheim, Structure and electronic properties of MoS_2 nanotubes, *Phys. Rev. Lett.* **85**, 146–49 (2000); b. A.N. Enyashin, S. Gemming, and G. Seifert, Nanosized allotropes of molybdenum disulfide, *Eur. Phys. J. Special Topics* **149**, 103–125 (2007).

[46] A.N. Enyashin, Theoretical studies of inorganic fullerenes and fullerene-like nanoparticles, *Isr. J. Chem.* **50**, 468–483 (2010).

[47] a. I. Kaplan-Ashiri, S.R. Cohen, K. Gartsman, V. Ivanovskaya, T. Heine, G. Seifert, I. Wiesel, H.D. Wagner, and R. Tenne, On the mechanical behavior of WS_2 nanotubes under axial tension and compression, *Proc. Natl. Acad. Sci.* **103**, 523–528 (2006); b. O. Tevet, P. Von-Hute, R. Popovitz-Biro, R. Rosentsveig, H.D. Wagner, and R. Tenne, Friction mechanism of individual multilayered nanoparticles, *Proc. Natl. Acad. Sci.* **108**, 19901–19906 (2011).

[48] a. L. Rapoport, Yu. Bilik, Y. Feldman, M. Homyonfer, S.R. Cohen, and R. Tenne, Hollow nanoparticles of WS_2 as potential solid-state lubricants, *Nature* **387**, 791–793 (1997); b. L. Yadgarov, R. Rosentsveig, G. Leitus, A. Albu-Yaron, A. Moshkovith, V. Perfilyev, R. Vasic, A.I. Frenkel, A.N. Enyashin, G. Seifert, L. Rapoport, and R. Tenne, Controlled doping of MS_2 (M = W, Mo) nanotubes and fullerene-like nanoparticles, *Angew. Chem. Int. Ed.* **51**, 1148–1151 (2012).

[49] L. Rapoport, N. Fleischer, and R. Tenne, Applications of WS_2 (MoS_2) inorganic nanotubes and fullerene-like nanoparticles for solid lubrication and for structural nanocomposites, *J. Mater. Chem.* **15**, 1782–1788 (2005).

[50] a. M. Chhowalla and G.A.J. Amaratunga, Thin films of fullerene-like MoS_2 nanoparticles with ultra-low friction and wear, *Nature* **407**, 164–167 (2000); b. F. Svahn and S. Csillag, Formation of low-friction particle/polymer composite tribofilms by tribopolymerization, *Tribol. Lett.* **41**, 387–393 (2011).

[51] a. A. Katz, M. Redlich, L. Rapoport, H.D. Wagner, and R. Tenne, Self-lubricating coatings containing fullerene-like WS_2 nanoparticles for orthodontic wires and other possible medical applications, *Tribol. Lett.* **21**, 135–139 (2006); b. A.R. Adini, Y. Feldman, S.R. Cohen, L. Rapoport, A. Moshkovith, M. Redlich, Y. Moshonov,

[51] (cont.) B. Shay, and R. Tenne, Alleviating incidental and fatigue-related failure of NiTi root canal files by self-lubricating coatings, *J. Mater. Res.* **26**, 1234–1242 (2011).

[52] M. Naffakh, C. Marco, M.A. Gomez, and I. Jimenez, Novel melt-processable nylon-6/inorganic fullerene-like WS$_2$ nanocomposites for critical applications, *Mater. Chem. Phys.* **129**, 641–648 (2011).

[53] a. M. Shneider, H. Dodiuk, S. Kenig, and R. Tenne, The effect of tungsten sulfide fullerene-like nanoparticles on the toughness of epoxy adhesives, *J. Adhes. Sci. Technol.* **24**, 1083–1095 (2010); b. E. Zohar, S. Baruch, M. Shneider, H. Dodiuk, S. Kenig, R. Tenne, and H.D. Wagner, The effect of WS$_2$ nanotubes on the properties of epoxy-based nanocomposites, *J. Adhes. Sci. Tech.* **25**, 1603–1617 (2011).

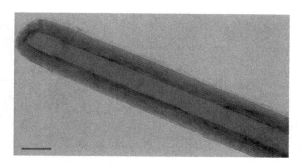

FIG. 3 Electron micrograph of part of a tubular crystal of WS_2. It is believed to be hollow; the residual contrast inside the tube may result from the outer wall perpendicular to the beam. Scale bar is 10 nm.

irregularity in the crystal structure or to the absence of crystalline material at these edges.

An alternative mechanism for the growth of curved structures may involve the participation of contaminating atoms. For example the structure of graphitic oxide is believed to consist of puckered carbon layers in which the carbon-carbon bonds are roughly tetrahedral[14]. This suggests the possibility that the strain of bending graphite (and WS_2) layers is reduced or eliminated by oxide (or some other 'contaminant') formation at the bends[9]. The source of the 'contaminants', such as oxygen species, could be due to diffusion out of the quartz at the high temperatures. Another growth mechanism may involve wrapping a crystalline core with sheets of layered compounds[15]. This growth mechanism seems to be irrelevant to the present case, where no crystalline core was observed.

The polyhedral WS_2 structures can join together to give chains or multiple structures, as is clear from the three connected crystals in Fig. 2c. The connection between the two crystals does not appear to be due to a simple van der Waals attraction, but rather a common crystal plane which undergoes bifurcation in two directions. This could occur by either of two pathways. In the first, two crystallites nucleate close to each other and grow until they merge, with the final sheet common to both of them. The second possibility is that nucleation begins at only one place, and the growing WS_2 sheet splits into two branches, a common occurrence, as we have previously shown (Fig. 7b in ref. 13). The two branches curl round in opposite directions, leading to two connected polyhedra. The bifurcation could occur at any stage in the growth of the initially nucleated crystal. An analogous mechanism has been suggested for the growth of adjacent graphitic spheres from amorphous carbon during high-energy electron irradiation[5]. □

Received 22 July; accepted 11 November 1992.

1. Kroto, H. W., Heath, J. R., O'Brien, Curl, R. F. & Smalley, R. E. *Nature* **318**, 162-163 (1985).
2. Iijima, S. *Nature* **354**, 56-58 (1991).
3. Iijima, S., Ichihasi, T. & Ando, Y. *Nature* **356**, 776-778 (1992).
4. Kroto, H. W. *Angew. Chem. Int. Edn Engl.* **31**, 111-129 (1992).
5. Ugarte, D. *Nature* **359**, 707-709 (1992).
6. Bates, T. F., Sand, L. B. & Mink, J. F. *Science* **111**, 512-513 (1950).
7. Yada, K. *Acta cryst.* **23**, 704-707 (1967).
8. Iijima, S. *J. Cryst. Growth* **50**, 675-683 (1980).
9. Bacon, R. *J. appl. Phys.* **31**, 283-290 (1960).
10. Pauling, L. *Proc. Natn. Acad. Sci. U.S.A.* **16**, 578-582 (1930).
11. Heidenreich, R. D., Hess, W. M. & Ban, L. L. *J. appl. Cryst.* **1**, 1-19 (1968).
12. Moser, J., Liao, H. & Lévy, F. *J. Phys. D. appl. Phys.* **23**, 624-626 (1990).
13. Genut, M., Margulis, L., Hodes, G. & Tenne, R. *Thin Solid Films* **217**, 91-97 (1992).
14. Aragon de la Cruz, F. & Cowly, J. M. *Acta cryst.* **16**, 531-534 (1963).
15. Bursill, L. A. *Int. J. mod. Phys.* **B4**, 2197-2216 (1990).
16. *Powder diffraction file*, Card 8-237 (ASTM, Philadelphia, Pennsylvania).

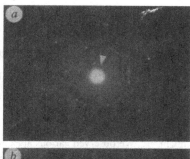

FIG. 4 Typical electron diffraction patterns of (a) a spheroidal crystal; the {00.2} ring is arrowed, the other spots belong to {hk.0} reflections; (b) a cylindrical crystal. The arrow shows the direction of the cylinder axis. All the reflections are elongated in the direction perpendicular to the cylinder axis, because of the needle-like shape of the cylinders.

Nested fullerene-like structures

SIR — Hollow carbon structures consisting of nested graphitic shells have been reported for various geometries[1-3], and the filling of these structures with metal compounds has also been described[4-6]. Our discovery[7] that similar hollow structures can be observed for the layered semiconductor tungsten disulphide leads one to expect that these structures might be formed from other layered materials. We have now found that molybdenum disulphide will also form such concentric structures, and that they too can be filled with metal compounds.

We deposited molybdenum films of 20 nm thickness onto quartz substrates, oxidized them at 500–600 °C in open air and fired them at 850–1,050 °C in a stream of H_2S and N_2/H_2 mixture. We peeled the MoS_2 films obtained off the substrate and examined them in the transmission electron microscope.

The figure (a) shows the lattice image of a few nested shells of MoS_2. The nested shells seem to be hollow in most cases. We also saw, albeit rarely, 'stuffed' structures, composed of molybdenum-dense core (confirmed by local energy dispersive X-ray analysis) wrapped with a few MoS_2 layers (inset in figure). X-ray photoelectron spectroscopy analysis of the films reveals a Mo:S atomic ratio of 1:2. This result indicates that the oxide serves as a precursor for the generation of a nested shell, rather than being a part of it.

We observed various apex angles, predominantly 120° and larger, but often of 60, 75 and 90°. Tilting about different axes shows that these angles are not necessarily symmetrical.

Some of the apex angles can be deduced from the MoS_2 layer structure. The figure (b, c, d) demonstrates how an apex can be created by folding and joining either three or four hexagons around a triangle or rhombus, respectively. Further growth of the cone may be explained by a model described by Kroto[8] for carbon fullerenes. Only one point defect — a molybdenum vacancy — is required for initiating this process. The generation of the 'stuffed' shells can be ascribed to a wrapping mechanism of spiral growth[9].

Structural transformations in the films

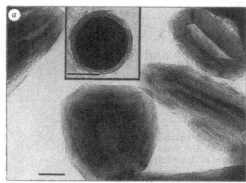
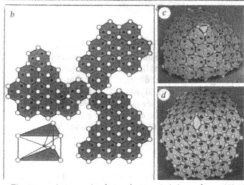

a, Electron micrograph of a region containing a few nested polyhedral shells. The interlayer distance in the different faces of the polyhedra is 0.62 nm, which coincides with the *c*/2 lattice parameter of 2H–MoS_2 polytype. A 'stuffed' Mo core with an incomplete MoS_2 shell is shown in the inset. Scale bars, 10 nm. *b*, Two-dimensional drawing of the Mo–S–Mo basic triple layer and a view of the elemental trigonal prism of 2H–MoS_2 lattice. Open circles, sulphur; closed circles, molybdenum. *c*, Three-dimensional drawing of a conical apex of 60° angle formed by folding and joining three hexagons around a triangle (as shown in *b*). *d*, Three-dimensional drawing of a conical asymmetric apex formed by folding and joining four hexagons around a rhombus, with 90° along the short axis of the rhombus and 75° along the long axis.

were studied by X-ray diffraction and absorption spectra. At 850 °C (30 min annealing time) most of the integrated intensity of the X-ray spectrum is confined to a wide peak of the amorphous material, and the (002) peak of the crystalline MoS_2 is very weak. The absorption spectrum does not show a particular structure. At 1,050 °C, on the other hand, a strong (002) peak of the crystalline 2H–MoS_2 phase predominates and clear excitonic peaks are observed at 680 and 630 nm (A and B excitons). Similar transformation of the diffraction peaks occurs when the annealing time (at 950 °C) increases from 5 min to 1 h. These results show that the amorphous material gradually transforms into a crystalline phase on annealing in H_2S atmosphere. The higher the temperature, the shorter is the time required for this transformation to occur.

At early stages of the annealing one observes in the electron microscope partially crystallized clusters, embedded in amorphous matrix. Tilting the sample at various angles inside the microscope reveals the incomplete structure of the nested shell. This stage is reminiscent of the closure of carbon polyhedra[8,10,11]. Annealing the films at 850 °C for 30 min or more yields a mixture of phases, containing round-shaped globules of about 100 nm or larger, and many hollow shells less than 20 nm in diameter.

The synthesis of MoS_2 from MoO_3 and sulphur (or H_2S) is well established[12]. The reaction leading from MoS_3 to MoS_2 is also known. A series of phase transformations occurs which can be represented by the sequence a:Mo → a:MoO_3 → a:MoS_3 → c:2H–MoS_2 (where 'a' means amorphous and 'c' crystalline). The nested-shell phase is an intermediate stage between the last two phases.

This and previous[7] work shows that closed polyhedra are a common phenomenon in materials with layered structure. One may now predict the discovery of nested shells in other layered materials: for example, our preliminary experiments confirm their presence in Mo(W) diselenides.

L. Margulis
G. Salitra
R. Tenne*
Department of Materials and Interfaces,
Weizmann Institute,
Rehovot 76100, Israel
M. Tallanker
Department of Materials Engineering,
Ben-Gurion University,
Beer Sheva 84105, Israel

* Author for correspondence.

1. Iijima, S. *J. cryst. Growth* **50**, 675–683 (1980).
2. Iijima, S. *Nature* **354**, 56–58 (1991).
3. Ugarte, D. *Nature* **359**, 707–709 (1992).
4. Ruoff, R. S. *et al. Science* **259**, 346–348 (1993).
5. Saito, Y. *et al. Jap. J. appl. Phys.* **32**, L280–L282 (1993).
6. Iijima, S. & Ajayan, P. M. *Nature* **361**, 333–334 (1993).
7. Tenne, R., Margulis, L., Genut, M. & Hodes, G. *Nature* **360**, 444–446 (1992).
8. Kroto, H. *Science* **242**, 1139–1145 (1988).
9. Bursill, L. *Int. J. mod. Phys.* B**4**, 2197–2216 (1990).
10. Heidenreich, R., Hess, W. & Ban, L. *J. appl. Crystallogr.* **1**, 1–19 (1968).
11. Curl, R. & Smalley, R. *Science* **242**, 1017–1022 (1988).
12. Zelikman, A., Chistyakov, Yu, Indenbaum, G. & Krein, O. *Soviet Phys. Crystallogr.* **6**, 308–312 (1961).

energy with respect to dislocation density yields [13,15]

$$E_b = \begin{cases} B_0 h^3, & \kappa < \kappa_c, \\ B_1 h + B_2 \sqrt{\nu h} - B_3 \nu/h^2, & \kappa > \kappa_c, \end{cases} \quad (2)$$

where $B_0 = \pi \chi c^3/3$, $B_1 = B_0$, $B_2 = B_0 \beta J/(\chi \pi^{1/2})$, $B_3 = B_0(\beta J)^2/(4\pi \chi^2)$, $\alpha = 1 - 1/h^2$, and $\beta = [3/(\pi \alpha)]\ln(R_0/r_0)$, and the critical curvature is defined by $\kappa_c = \beta J/(\chi h^2 c)$ [13]. In the derivation of Eq. (2), we assumed that although the shell is thin compared to the radius $h/r \ll 1$, it is thick compared to an atomic spacing $h \gg 1$, and terms of higher order than linear in h/r are neglected. For $\kappa < \kappa_c$, the total elastic energies of nested fullerenes are lower when they are bent coherently (without dislocations). However, for $\kappa > \kappa_c$, the elastic energy is lowered by forming a dislocation array of constant density. Therefore, a phase transition between undislocated and dislocated, nested fullerenes occur at a finite, critical curvature κ_c, which is dependent on the elastic constant of the material and is inversely proportional to the square of the thickness. For a fixed curvature, one can think of a critical thickness below which the layers bend with no dislocations and above which the structure contains dislocations. The theory shows that thin, nested fullerenes have high critical curvatures and one might expect them to grow relatively dislocation free. The same analysis and conclusions apply to the formation of grain boundaries or polygonalized morphologies; dislocations tend to organize themselves into grain boundaries and theory [13] predicts that thin, nested fullerenes would tend to be grain boundary free and show spherically symmetric morphologies.

In addition to the bulk terms, the total energy of a nested fullerene also includes contributions from the inner and outer surfaces of the spherical shell. In all known nested fullerenes, the bonding within the atomic layers (i.e., the basal planes) is covalent, while that between layers is primarily due to van der Waals forces. Therefore, the surface energy can be written $E_s = S\nu/h$, where $S = 2\gamma c^2$ and γ is the surface energy (per unit area) of a semi-infinite solid of the film material and related to the van der Waals energies.

The total energy of the nested fullerene is $E_{tot} = E_s + E_b$. The equilibrium nested fullerene thickness may be obtained by minimizing E_{tot} with respect to h at fixed ν. The balance between the h^3 and $1/h$ terms determine the equilibrium aspect ratio $\eta = h/r = d/R$ for $\kappa < \kappa_c$. The large ν limit, the competition between the terms proportional to $h^{1/2}$ and $1/h$ determine η_{eq}, when $\kappa > \kappa_c$. For large ν, the equilibrium aspect ratio η_{eq} is

$$\eta_{eq} \approx \begin{cases} (2\sqrt{\pi}/3^{3/8})(S^3/B_0^3 \nu)^{1/8}, & \kappa < \kappa_c, \\ (4\sqrt{\pi})(S/B_2), & \kappa > \kappa_c, \end{cases} \quad (3)$$

Equation (3) suggests that the equilibrium thickness to radius ratio of nested fullerenes should decrease slowly with an increasing number of atoms when $\kappa < \kappa_c$, i.e., small curvature and/or thickness. On the other hand,
when $\kappa > \kappa_c$ (large curvature and/or thickness), η_{eq} is constant to leading order in ν, and is determined by the surface energy to relaxed bending energy ratio.

The equilibrium shape of nested fullerenes predicted by Eq. (3) is evaluated using the accurately known elastic constants [16] of graphite; the uncertainty in the (0001) surface energy is of the order of 50% [17]. In MoS$_2$, the elastic constants are also known [18], but the uncertainty in the (0001) surface energy is of the order of a factor of 10 [19,20]. Using the available data, we estimate that the equilibrium aspect ratio is $\eta = 0.06$ for the graphitic, nested fullerene and $\eta = 0.1$ for MoS$_2$, assuming that $\kappa > \kappa_c$. The qualitative message is that *equilibrium* large scale fullerenes should be "thin," i.e., they should have a small thickness to radius ratio.

Similar considerations can be used to analyze the energies of polygonalized, nested fullerenes, where the polygonalization is treated in terms of grain boundaries. Speck [21] was the first to compare the energetic's spherical and faceted carbon layers in a study of carbon blacks, but did not consider the important effects of dislocations or anisotropy or determine the equilibrium value of η which are included in the present analysis. The energy associated with the grain boundaries in a nested fullerene is equal to the grain boundary energy per unit area γ_{gb} times the total grain boundary area. The grain boundary area is proportional to h and the total length of all edges, which is proportional to R. The proportionality constant ξ depends on the type of polyhedron and typically decreases slowly with an increasing number of faces for fixed volume. The total energy of a faceted, nested fullerene has contributions from both the grain boundary energy ($E_{gb} = \gamma_{gb} h R \xi$) and the energies of the inner and outer surfaces. Following the same procedure used to determine the total energy of dislocated, nested fullerenes, we find $E_{tot}^{gb} = G\sqrt{\nu h} + S\nu/h$ for $\kappa > \kappa_c^{gb}$, where $G = \xi \gamma_{gb} c^2/\sqrt{4\pi}$ and κ_c^{gb} are the critical curvature for forming grain boundaries. In isotropic materials, $\kappa_c^{gb} < \kappa_c$ [13]. For $\kappa < \kappa_c^{gb}$, the energy is as given in Eq. (2). E_{tot}^{gb} exhibits a minimum at $\eta_{eq}^{gb} = h/r = d/R = 4\sqrt{\pi}S/G$. Therefore, η_{eq}^{gb} is independent of the number of atoms ν in the large ν limit and is determined by the ratio of the surface to the relaxed bending energies.

Comparison of η_{eq}^{gb} with η_{eq}, shows that the η_{eq} for the two relaxation mechanisms are both independent of ν and proportional to the surface energy S. The ratio $\eta_{eq}^{gb}/\eta_{eq} \approx [2cJ\ln(R_0/r_0)]/(\xi \gamma_{gb}\alpha)$ is approximately 5 for graphite. The uncertainty in this value of η_{eq}^{gb} is of the same order as that for η_{eq}, owing to the uncertainty in the grain boundary energy. This value of the ratio is based upon the assumption that the polyhedral, nested fullerene is icosahedral such that $\xi \approx 20$, $J = 6.5 \times 10^{10}$ Pa, $\gamma_{gb} = 1$ J/m^2, $\alpha = 1$, $c = 2 \times 10^{-10}$ m, and $\ln(R_0/r_0) = 5$. In any case, we expect that $\eta_{eq}^{gb} > \eta_{eq}$, since the dislocation energy is typically lowered when they organize into grain boundaries. This would also suggest that polyhedral (faceted), nested fullerenes should have lower energy than those relaxed by dislocations (for the same ν). However, since there are no topological constraints on the number of grain boundaries and the grain boundary energy changes rapidly with misorientation at small angles, it is possible for dislocation relaxation to lead to lower energy structures as compared with structures with many low-angle grain boundaries. The rearrangement of dislocations into a relatively small number of high misorientation grain boundaries may be limited by kinetics.

One obvious difficulty in experimentally determining η is the necessity to distinguish between shapes determined by growth and/or kinetic factors and true equilibrium shapes. Clearly, if the nested fullerenes grow in a layer-by-layer mode, then it is unlikely that the resultant structures are in equilibrium. Carbon, nested fullerenes are typically found to have interior radii very close to that of C_{60}, while the outer radii vary over a substantial range [22]. Nested fullerenes of MoS_2, on the other hand, can show a wide range of both inner and outer radii (see Fig. 1). Typical MoS_2, nested fullerene, produced from the sulfidation of a very thin film of metal oxide on a quartz substrate, is shown in the atomic resolution electron micrograph in Figs. 1(a) and 1(b). Examination of the nearly spherical, nested fullerene in Fig. 1(a) shows that such structures may be described in terms of a relatively large number of low-angle grain boundaries or by a uniform array of dislocations. The fullerene in Fig. 1(b) is strongly faceted, with very flat atomic planes meeting at sharply defined grain boundaries. We note that both of these samples are grown under conditions which are far from equilibrium and comparison with the theory is not possible.

Recently, however, a new technique for the gas phase synthesis of MoS_2, nested fullerenes has been developed [23,24]. In this technique, which affords a much better control over the growth conditions than the previous gas-solid synthetic route [4], gaseous MoO_{3-x} and H_2S gases diluted in a carrier gas are reacted at elevated temperatures ($>800\,°C$). It is expected that, in this synthetic route, the nested fullerenes grow independently of each other, mostly through interaction with the carrier (forming) gas and hence growth conditions which are closer to equilibrium prevail. Copious amounts of the nested fullerene and nanotube phases are obtained in this way.

It appears that even in this synthesis the role of the substrate, on which the nested fullerenes are being collected, cannot be overlooked [23,24]. If a NbS_2 substrate is used, large amounts of nested fullerenes consisting of two shells and of various shapes are observed. A typical example of a nested fullerene with $\eta = 0.2$ and a nearly spherical shape is shown in Fig. 1(c). Most of the nested fullerenes produced in this way exhibit values of $\eta \leq 0.2$, the majority of these are at least partially faceted—containing at least one grain boundary. In other

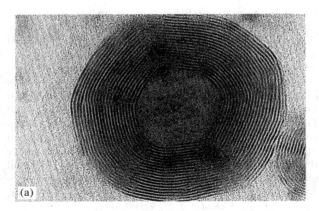

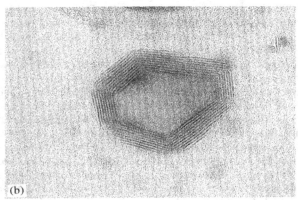

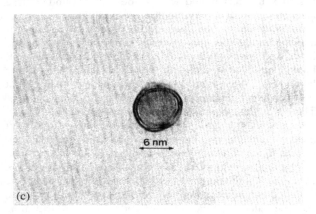

FIG. 1. High resolution transmission electron microscope image of MoS_2 fullerenes. The distance between the fringes is 0.62 nm, which is half of the c-axis lattice constant in hexagonal MoS_2. (a) and (b) show nested MoS_2 fullerenes prepared by firing a 10 nm film of MoO_3 in H_2S under reducing conditions (see [4] for more details) with (a) a nearly spherical shape and (b) a polyhedral, faceted shape, respectively. (c) shows a nearly spherical, two-layer, nested fullerene produced in a gas phase reaction between a stream of gaseous molybdenum suboxide and H_2S diluted in a carrier gas of H_2/N_2 (5%/95%).

cases, where amorphous carbon has been used as a substrate, single-layer nonfaceted fullerenes with a spherosymmetric shape were recently obtained [24]. In another reactor of different design with a quartz substrate as a collector [23], values of η closer to 0.7 were typically obtained. It is very likely that under the harsh flow conditions in this reactor, the lighter, nested fullerenes (with $\eta = 0.1-0.2$) are swept away by the gas flow and are collected on some other parts of the reactor farther upstream from the substrate. Thus, although a conclusive comparison between theory and experiment cannot be made at this time, we now have experimental evidence that large quantities of nested fullerenes with η as small as 0.1 do occur. These experimental results are in agreement with our theoretical predictions presented above.

The HRTEM micrographs were prepared in collaboration with Dr. J. L. Hutchinson and Dr. L. Margulis. D.I.S. gratefully acknowledges the hospitality of The Weizmann Institute of Science and the support of the Division of Materials Science of the U.S. Department of Energy (Grant No. FG02-88ER-45367). S.A.S. is grateful for the support of the Donors of the Petroleum Research Fund, administered by the ACS and the Israel Academy of Sciences. M.H. and R.T.'s research was sponsored by the US-Israel Binational Science Foundation and the Manof project of the Israeli Ministry of Science and Arts.

---

*Permanent address: Dept. of Materials Science and Engineering, University of Michigan, Ann Arbor, MI 41809-2136.

[1] D. Ugarte, Europhys. Lett. **22**, 45 (1993); Nature (London) **359**, 707 (1992).
[2] A. Maiti, C. J. Brabec, and J. Bernholc, Phys. Rev. Lett. **70**, 3023 (1993).
[3] J. P. Lu and W. Yang, Phys. Rev. B **49**, 11 421 (1994).
[4] R. Tenne, L. Margulis, M. Genut, and G. Hodes, Nature (London) **360**, 444 (1992); L. Margulis, G. Salitra, M. Talianker, and R. Tenne, Nature (London) **365**, 113 (1993); M. Hershfinkel, L. A. Gheber, V. Volterra, J. L. Hutchinson, L. Margulis, and R. Tenne, J. Am. Chem. Soc. **116**, 1914 (1994).
[5] J. Tersoff, Phys. Rev. B **46**, 15 546 (1992).
[6] G. B. Adams, O. F. Sankey, J. B. Page, M. O'Keeffe, and D. A. Drabold, Science **256**, 1792 (1992).
[7] Q.-M. Zhang, J.-Y. Yi, and J. Bernholc, Phys. Rev. Lett. **66**, 2633 (1991).
[8] B. L. Zhang, C. H. Xu, C. Z. Wang, C. T. Chan, and K. M. Ho, Phys. Rev. B **46**, 7333 (1992).
[9] D. Tomanek and M. A. Schluter, Phys. Rev. Lett. **67**, 2331 (1991).
[10] P. Ballone and P. Milani, Phys. Rev. B **42**, 3201 (1990).
[11] L. D. Landau and E. M. Lifshitz, *Theory of Elasticity* (Pergamon, Oxford, 1986), 3rd ed.
[12] W. Helfrich, Z. Naturforsch. **28c**, 693 (1973).
[13] D. J. Srolovitz, S. A. Safran, and R. Tenne, Phys. Rev. E **49**, 5260 (1994).
[14] J. P. Hirth and J. Lothe, *Theory of Dislocations* (John Wiley, New York, 1982).
[15] Tersoff [5] showed that for a single layer of curved graphite, the bending energy has logarithmic corrections (i.e., $\log R$) attributable to the disclinations that must be present in order to fold a sheet into a closed, curved shape. This generates additional logarithmic terms in Eq. (2) which rescale the constants in this equation but are much less important than the algebraic terms in determining the morphology.
[16] O. L. Blakeslee, D. G. Proctor, E. J. Seldin, G. B. Spence, and T. Weng, J. Appl. Phys. **41**, 3373 (1970).
[17] J. Abrahamson, Carbon **11**, 337 (1973).
[18] J. L. Feldman, J. Phys. Chem. Solids **37**, 1141 (1976).
[19] M. A. Berding, S. Krishnamurthy, A. Sher, and A.-B. Chen, J. Appl. Phys. **67**, 6175 (1990).
[20] A. I. Brudnyi and A. F. Karmadonov, Wear **33**, 243 (1975).
[21] J. S. Speck, J. Appl. Phys. **67**, 495 (1990).
[22] S. Iijima, J. Cryst. Growth **50**, 675 (1980).
[23] Y. Feldmann, E. Wasserman, D. Srolovitz, and R. Tenne, Science (to be published).
[24] M. Homyonfer and R. Tenne (to be published).

Doped and Heteroatom-Containing Fullerene-like Structures and Nanotubes**

By Reshef Tenne*

A review of the current knowledge on carbon fullerenes doped with foreign atoms and heteroatom-containing fullerene-like structures and nanotubes is provided. Strong covalent bonds lend high specificity and stabilize the hollow-cage and symmetric structure of these moieties. These structures are distinct from noble-gas and metallic clusters, where either weak van der Waals forces or stronger metallic bonds, which are not very specific, hold the cluster atoms together. In the latter kind of clusters, atoms gain stability through close packing and large coordination numbers and consequently they cannot afford a hollow core. Nonetheless some intermetallic nanoparticles exhibit truncated icosahedral symmetry. The field is divided, somewhat artificially, into two separate categories. One family consists of fullerene-like clusters assembled from different atoms which do not have a bulk counterpart of similar chemical formula. The other group is that of fullerene-like nanostructures which are obtained mainly from ubiquitous 2-D layered compounds; various elements and compounds with 3-D character and also from certain metallic alloys. It is shown that nanoparticles of 2-D compounds are unstable in the planar form and they reconstruct into hollow-cage nanoparticles, spontaneously. Nanosolids of this kind may reveal vastly different properties from their bulk predecessors. Numerous applications for the doped and heteroatom fullerene-like materials in the fields of catalysis, lubrication, electronic and photonic devices, alternative energy sources, etc. are expected upon further study and development.

1. Introduction

Graphite is the most stable form of carbon under ambient conditions. Notwithstanding, graphite nanoparticles have been shown to be unstable and they close into fullerenes,[1] nested fullerenes[2] and nanotubes.[3] The stimulus for graphite nanoparticles to spontaneously form carbon fullerenes (CF) is believed to emanate[4] from the large energy associated with the dangling bonds of the peripheral sp^2 bonded carbon atoms. The extreme stability of C_{60} was attributed to the fact that this is the smallest fullerene with disjoint pentagons (the so-called *pentagon rule*).[4,5] The discovery of carbon fullerenes and nanotubes with truncated icosahedral symmetry and the means to produce them in large quantities,[6,7] has instigated a surge of research in this field. This work has been discussed in a number of reviews (e.g. see [8]), and up-to-date proceedings volumes of conferences are frequently published.[9] Hence, this subject will not be further covered in the present review article.

Endohedral fullerenes, in which a foreign atom is inserted into the hollow core of the carbon cage, were first reported shortly after the confirmation of the C_{60} structure.[10] In the case of C_{60} a calculation shows that small atoms and alkali-metal cations like H, He, Ne, $Li^\ominus$ and $Na^\oplus$ occupy an off-center site, while large atoms and alkali-metal ions like Ne, Kr and $K^\oplus$, occupy an on-center site within the hollow cage.[11] Larger fullerenes, were shown to admit more than one atom within the cage, e.g. $La_2@C_{80}$[12] and $Sc_3@C_{82}$.[13] In many cases the encapsulated atom is not held merely through weak van der Waals (vdW) forces but rather, electron transfer from the metal atom to the carbon cage results in a strong ionic bond between the positively charged metal ion and the negatively charged carbon cage.

Yet another important discovery was that of nanotubes which were stuffed with metal or metal carbides, like LaC_2[14] and YC_2.[15] Strikingly, encapsulated Co, Fe and Ni nanoparticles can serve as a kind of a catalyst which spearheads the growth of a single-layer nanotube.[16] The topic of endohedral fullerenes and nanotube encapsulation by metal (metal carbide) has been expanded in the meantime to such an extent that it would be impossible to provide a concise progress report within the framework of this review article.

Fullerenes serve as a paradigm for the ability of clusters to adopt crystalline structure which is radically different from that of bulk materials. Further studies have led to the observation[17,18] that in a reaction between carbon and transition metal atoms (M), such as Ti, V, Hf, Zr, etc., a new class of metal carbohedranes (Met-Cars) clusters of the formula $C_{12}M_8$, are obtained. These clusters consist of twelve 5-membered rings. The exchange of 8 carbon atoms by foreign metal atoms stabilizes a cluster with dodecahedron symme-

---

[*] Prof. R. Tenne
Department of Materials and Interfaces
The Weizmann Institute of Science
Rehovot 76100 (Israel)

[**] I am indebted to my collaborators and students: Dr. L. Margulis, Prof. S. A. Safran, Prof. D. J. Srolovitz, Dr. E. Wasserman, Dr. J. L. Hutchison, Dr. G. Hodes, Prof. C. Colliex, M. Homyonfer, Y. Feldman and G. Frey. The support of the following organizations and agencies is acknowledged: The Edith Reich and the Levine foundations of the Weizmann Institute, the US–Israel binational science foundation, Israeli Ministry of Science and Arts (MANOF project), and Ministry of Energy and Infrastructure.

electrons pointing outwards from the sp^3 bonded Si[74] and also to the diamond surface.

An example of a class of fullerene-like materials with sp^3 bonded carbon are the collapsed fullerites,[75, 76] like C_{100} and hydrogenated dodecahedrane $C_{20}H_{20}$. C_{100} consists of an inner C_{20} layer (Fig. 14) with a hollow core, having sp^3

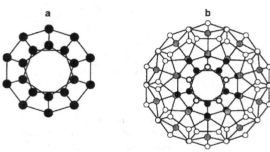

Fig. 14. Schematic structure of the sp^3 bonded C_{100} fullerene. Adopted with changes from [76].

bonded atoms arranged in icosahedral symmetry. A cover layer (C_{20}), which saturates the "surface" dangling bonds of the inner structure comes next, and finally a top layer of C_{60} at the outer surface of the cluster. In the fullerene-like $C_{20}H_{20}$, the hydrogen atoms saturate the "surface" dangling bonds. It is not unlikely that other elements, in particular Si atoms could form stable clusters of similar topology.

Saturation of the bonds of the C atoms which point outside the cage of C_{20} is not limited to hydrogen alone. Fluorine atoms were shown to function in much the same way.[77] Dodecahedral clusters of 20 sp^3 bonded atoms with icosahedral symmetry (I_h) were proposed for C, Si, Ge.[78, 79] A transition from the elongated puckered structure for $n < 27$ to spherical clusters for a larger number of Si atoms was observed[80] and substantiated by theoretical calculations.[26]

The internal energy of Si_{60} and Ge_{60} fullerene-like structures was calculated. They were both found to have appreciably inferior stability relative to C_{60}.[81] Stable Si_7 clusters with pentagonal bipyramid (D_{5h} symmetry) structure were trapped in a solid N_2 matrix (15 K) and identified by Raman measurements.[82] The reactivity of the Si clusters was shown to exhibit an oscillatory behavior,[83] which indicated the existence of "magic numbers" for such clusters as Si_{12} and Si_{45}. A recent numerical analysis[84] attributes the extra stability of such clusters to closed-shell fullerene-like structures. Thus, although sp^3-bonded atoms can form isolated clusters with fullerene-like structures, their stability seem to be a far-cry from that of C_{60} and the like.

Foremost among the elements that pack in icosahedral (and other Archimedean) structures is boron, its various hydrides (boranes) and related boron compounds,[67] a topic which has been covered by numerous reviews and books and will not be further discussed in this review.

Some careful considerations are required to exploit the concept of fullerene-like structures in covalently bonded binary AB compounds with a 3-D nature (like GaAs). The strength of the A–B chemical bond entails a disposition of the atoms such that A–A or B–B nearest neighbors (chemical frustration) are prohibited. Therefore, no rings with an even number of atoms are expected and the inner core of the fullerene-like structure can be expected to be akin to e.g. $B_{12}N_{12}$ (Fig. 11). Each atom of the inner core has an extra bond pointing out of the cage, which could be made to interact with an outer AB layer.

3.5.2. Connected Polyhedra

Since one out of four sp^3 bonds of a given atom is pointing out of the cage, the most stable fullerene-like structure in this case is a network of connected cages. This kind of network is realized in alkali-metal doped silicon clathrates,[25] which were found to have a fullerene-like structure.[85] In these compounds Si polyhedra of twelve 5-membered rings and two, or four more 6-membered rings share faces, and form a network of hollow cage structures which can accommodate endohedral metal atoms. Recently, the clathrate compound $(Na,Ba)_xSi_{46}$ has been synthesized (see Fig. 15) and demon-

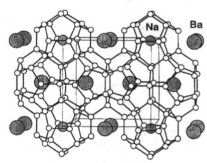

Fig. 15. Schematic illustration of the clathrate molecule $Na_2Ba_6Si_{46}$. Adopted with changes from [87].

strated a superconductor transition at 4 K.[86, 87] The electronic structures of Si_{46} clathrates were calculated[88] and found to differ markedly from that of bulk Si with the diamond structure. In particular, hybridization of Ba states in $Na_2Ba_6Si_{46}$ produces a very high density of states near the Fermi level which may explain the reported superconductivity.[86]

The possible implications of Si or Ge nano-solids for nanoelectronics cannot be overestimated, since the materials which will shape the electronic industry of tomorrow, once lithographic techniques are exhausted, can only be imagined at this stage. The observation of luminescence from nanocrystalline Si particles[89] (porous silicon) suggests that size-selective optoelectronics with nanoparticles of Si may be an attractive future technology.

3.6. Metallic Alloys with a Fullerene-like Structure

A few recent and intriguing reports suggest that metallic alloys with "polymerized" fullerene-like structures may occur.[90] In this work metal alloys with a composition akin to Zintl[91] phases, were shown to exhibit a complex structure consisting of a layer of interconnected fullerene-like cages which are "stuffed" with $In_{10}Ni^{\ominus}$ anionic clusters and Na polyhedra. Two phases with interconnected metallic fullerene structures were discovered: hexagonal $Na_{96}In_{97}Z_2$ (with Z = Ni, Pd, or Pt), and orthorhombic $Na_{172}In_{197}Z_2$. According to fullerene nomenclature each endohedral unit can be described as: $Ni@In_{10}@Na_{39}@In_{74}$. This unit contains the fullerene In_{74} (Fig. 16a) as the outer sphere. Another unit, which was found in both phases, can be designated as $Ni@In_{10}@Na_{32}@Na_{12}In_{48}$ and contains the fullerene cage $In_{48}Na_{12}$ (Fig. 16b). Various phases of intermetallics

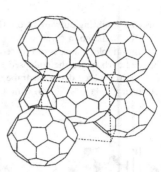

Fig. 17. Schematic bcc structure of the truncated icosahedra found in nanocrystallites of amorphous R-Al_5Li_3Cu. The dashed line shows the bcc unit cell. Adopted with changes from [93].

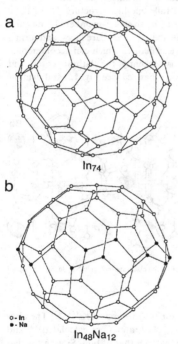

Fig. 16. Two polyhedra which make up part of the Zintl-like fullerenes: a) In_{74} (D_{3d} symmetry) in $Ni@In_{10}@Na_{39}@In_{74}$. b) $In_{48}Na_{12}$ (D_{3d} symmetry) in $Ni@In_{10}@Na_{32}@Na_{12}In_{48}$. Adopted with changes from [90].

and a few oxides, having a fullerene-like cage-structure, have also been reported.[92]

The pressure-induced transition[93] of crystalline intermetallic R-Al_5Li_3Cu into the amorphous Frank–Kaspar phase,[94] which consists of nanocrystallites of face-sharing units with truncated icosahedron structure (Fig. 17), has been reported. This study also shows the analogy between the quasicrystalline i-Al_6Li_3Cu phase with five-fold (icosahedral) symmetry,[95] which is obtained by rapid quench of the melt and the amorphous phase of R-Al_5Cu_3Li consisting of nanocrystallites of truncated icosahedral symmetry. They are both formed under kinetic control, where metastable phases cannot be excluded.

Notwithstanding the soft nature of the metallic bond, metals afford thermodynamic stability (and high electrical conductivity) by having a high coordination number for the atoms. In correspondence with these criteria metallic alloys crystallize in close-packed (fcc or hcp) lattice, where the maximum possible coordination number for a bulk atom (12) is attained, or body centered cubic (bcc) lattice, where the coordination number of a bulk atom is 8. The icosahedral packing in nanoparticles of metals and metallic alloys is no exception to that. The coordination number of 12 is achieved through "stuffing" the icosahedral hollow core with metallic atoms. Therefore, the metallic fullerene-like nanostructures do not have a hollow core, and could be thought of as being analogous to endohedral fullerenes in covalent materials.

Considering the fact that large fraction of the atoms are situated at the nanoparticle surface, the average coordination number of the atoms in a nanoparticle is smaller than 12. The difference in electronegativity between the atoms comprising the fullerene-skeleton (In_{74}) and the remainder of the metal atoms inside the core (Na_{39}), cause partial charging of the fullerene. This effect, which is further extenuated by the -10 charge of the inner $In_{10}Ni$ core, contributes to the extra stability of these polymeric structures, as much as electron transfer from the endohedral metal atom to the carbon cage stabilizes the endohedral fullerene cage.

4. Conclusions

The chemical variation of the fullerenes has been extended by the doping of the carbon network of C_{60} and similar materials with heteroatoms and especially by the discovery of inorganic fullerene-like structures formed from inorganic layered compounds. It is expected that within the next few years many more compounds belonging to this class of nanomaterials will be isolated and studied. The main issue is the understanding of the structure of these nanoclusters through molecular models. A far better control of the synthesis of these materials must be attained. Experimental and theoretical tools, which are capable of analyzing heavy clusters of that kind, must be developed. Such techniques may allow one to isolate the IF structures which are analogous to C_{60} and the larger fullerenes. The properties of these nanoparticles may reveal a plethora of new phenomena, unexplored so far, which may open up new applications for such nanomaterials.

The common ground to all the nanomaterials described in this review is the fact that they can exist only in nanophases, i.e. they are obtained under conditions where the corresponding bulk materials are either unstable or their formation is kinetically hindered. In such phases the nanocluster or nanocrystallite is a unit of its own, devoid of neighbors, and has no means to release its extra internal energy through further crystallization, but rather through interaction with the walls of the container or with the carrier gas.

The theoretical treatment of this class of materials and the dynamic processes which lead to their generation is not well developed at this point and an effort in this direction is warranted. The diversity of compounds which form fullerene-like structures is astounding. It is not easy to vindicate the search for a common principle in their formation, although this goal may be achieved through further research.

Received: April 7, 1995
Final version: August 1, 1995

[1] S. Iijima, *J. Cryst. Growth* **1980**, *50*, 675.
[2] H. W. Kroto, J. R. Heath, S. C. O'Brien, R. F. Curl, R. E. Smalley, *Nature* **1985**, *318*, 162.
[3] S. Iijima, *Nature* **1991**, *354*, 56.
[4] H. W. Kroto, *Nature* **1987**, *329*, 529.
[5] Q. L. Zhang, S. C. O'Brien, J. R. Heath, Y. Liu, R. F. Curl, H. W. Kroto, R. E. Smalley, *J. Phys. Chem.* **1986**, *90*, 525.
[6] a) W. A. Krätschmer, L. D. Lamb, K. Fostiropoulos, D. R. Huffman, *Nature* **1990**, *347*, 354. b) T. Guo, C. Jin, R. E. Smalley, *J. Phys. Chem.* **1991**, *95*, 4948.
[7] T. W. Ebbesen P. M. Ajayan, *Nature* **1992**, *358*, 220.
[8] a) *Mater. Res. Soc. Bull.* **1994**, *19*, No.11. b) G. S. Hammond V. J. Kuck (Eds.) *Fullerenes*, ACS Symp. Ser. **1992**, 481.
[9] Over the last few years the American Physical Society, Materials Research Society, the Electrochemical Society have all sponsored at least one annual symposium on fullerenes, which have resulted in a number of proceedings volumes. Other societies also organize symposia on fullerenes, almost annually.
[10] a) J. R. Heath, S. C. O'Brien, Q. Zhang, Y. Liu, R. F. Curl, H. W. Kroto, F. K. Tittel, R. E. Smalley, *J. Am. Chem. Soc.* **1985**, *107*, 7779. b) Y. Chai, T. Guo, C. Jin, R. E. Haufler, L. P. F. Chibante, J. Fure, L. Wang, J. M. Alford, R. E. Smalley, *J. Phys. Chem.* **1991**, *95*, 7564.
[11] J. Bretón, J. González-Platas, C. Girardet, *J. Chem. Phys.* **1993**, *99*, 4036.
[12] M. M. Alvarez, E. G. Gillan, K. Holczer, R. B. Kaner, K. S. Min, R. L. Whetten, *J. Phys. Chem.* **1991**, *95*, 10561.
[13] a) H. Shinohara, H. Sato, M. Ohkohchi, Y. Ando, T. Kodama, T. Shida, T. Kato, Y. Saito, *Nature* **1992**, *357*, 52. b) C. S. Yannoni et al., *Science* **1992**, *256*, 1191.
[14] a) R. S. Ruoff, D. C. Lorents, B. Chan, R. Malhorta, S. Subramoney, *Science* **1993**, *259*, 346. b) M. Tomita, Y. Saito, T. Hayashi, *Jpn. J. Appl. Phys.* **1993**, *32*, L280.
[15] S. Seraphin, D. Zhou, J. Jiao, J. C. Withers, R. Lotfy, *Nature* **1993**, *362*, 503.
[16] a) S. Iijima, T. Ichihashi, *Nature* **1993**, *361*, 603. b) D. S. Bethune, C. H. Kiang, M. S. deVries, G. Gorman, R. Savoy, J. Vazques, R. Beyers, *Nature* **1993**, *363*, 605.

[17] a) B. C. Guo, K. P. Kerns, A. W. Catleman, Jr., *Science* **1992**, *255*, 1411. b) B. C. Guo, S. Wei, J. Purnell, S. Buzza, A. W. Catleman, Jr., *Science* **1992**, *256*, 515. c) S. F. Cartier, Z. Y. Chen, G. J. Walder, C. R. Sleppy, A. W. Catleman, Jr., *Science* **1993**, *260*, 195.
[18] A. F. Wells, *Structural Inorganic Chemistry*, 5th ed., Clarendon Press, Oxford **1993**, pp. 947–953.
[19] M. L. Cohen, *Solid State Commun.* **1994**, *92*, 45.
[20] K. Kobayashi, N. Kurita, *Phys. Rev. Lett.* **1993**, *70*, 3542.
[21] a) O. Stephan, P. M. Ajayan, C. Colliex, P. Redlich, J. M. Lambert, P. Bernier, P. Lefin, *Science* **1994**, *266*, 1683. b) Z. Weng-Sieh, K. Cherrey, N. G. Chopra, X. Blase, Y. Miyamoto, A. Rubio, M. L. Cohen, S. G. Louie, A. Zettl, R. Gronsky, *Phys. Rev. B* **1994**, *51*, 11229. c) N. G. Chopra, R. J. Luyken, K. Cherey, V. H. Crespi, M. L. Cohen, S. G. Louie, A. Zettl, *Science* **1995**, *269*, 966.
[22] R. Tenne, L. Margulis, M. Genut, R. Tenne, *Nature* **1992**, *360*, 444.
[23] R. Tenne, L. Margulis, G. Hodes, *Adv. Mater.* **1993**, *5*, 386.
[24] L. Margulis, G. Salitra, R. Tenne, M. Talianker, *Nature* **1993**, *365*, 113.
[25] J. S. Kasper, P. Hagenmuller, M. Pouchard, C. Cros, *Science* **1965**, *150*, 1713.
[26] Kaxiriaks, K. Jackson, *Phys. Rev. Lett.* **1993**, *71*, 727.
[27] a) S. Ino, S. Ogawa, *J. Phys. Soc. Jpn.* **1967**, *22*, 1365. b) B. M. Smirnov, *Chem. Phys. Lett.* **1995**, *232*, 395.
[28] a) S. Wei, B. C. Guo, J. Purnell, S. Buzza, A. W. Castleman, Jr., *J. Phys. Chem.* **1992**, *96*, 4166. b) B. C. Guo, S. Wei, Z. Chen, K. P. Kerns, J. Purnell, S. Buzza, A. W. Castleman, Jr, *J. Chem. Phys.* **1992**, *97*, 5243.
[29] S. Wei, B. C. Guo, H. T. Deng, K. Kerns, J. Purnell, S. A. Buzza, A. W. Castleman Jr., *J. Am. Chem. Soc.* **1994**, *116*, 4475.
[30] J. S. Pilgrim, M. A. Duncan, *J. Am. Chem. Soc.* **1993**, *115*, 4395.
[31] J. S. Pilgrim, M. A. Duncan, *J. Am. Chem. Soc.* **1993**, *115*, 6958.
[32] a) B. V. Reddy, S. N. Khanna, P. Jena, *Science* **1992**, *258*, 1640. b) M. Methfessel, M. van Schilfgaarde, M. Scheffler, *Phys. Rev. Lett.* **1993**, *70*, 29.
[33] a) I. Dance, *J. Chem. Soc., Chem. Commun.* **1992**, 1779. b) H. Chen, M. Feyereisen, X. P. Long, G. Fitzgerald, *Phys. Rev. Lett.* **1993**, *71*, 1732.
[34] S. Lee, N. G. Gotts, G. von Helden, M. T. Bowers, *Science* **1995**, *267*, 999.
[35] a) B. V. Reddy, S. N. Khanna, *Chem. Phys. Lett.* **1993**, *209*, 104. b) B. V. Reddy, S. N. Khanna, *J. Phys. Chem.* **1994** *98*, 9446.
[36] S. F. Cartier, B. D. May, A. W. Castleman, Jr., *J. Phys. Chem.* **1994**, *100*, 5384.
[37] S. Wei, B. C. Guo, J. Purnell, S. A. Buzza, A. W. Castleman, Jr., *Science* **1992**, *256*, 818.
[38] S. Wei, A. W. Castleman, Jr., *Chem. Phys. Lett.* **1994**, *227*, 305.
[39] S. N. Khanna, *Phys. Rev. B* **1995**, *51*, 10965.
[40] H. T. Deng, B. C. Guo, K. P. Kerns, W. A. Castleman, Jr., *J. Phys. Chem.* **1994**, *98*, 13373.
[41] D. E. Clemmer, J. M. Hunter, K. B. Shelimov, M. F. Jarrold, *Nature* **1994**, *372*, 248.
[42] M. Hershfinkel, L. A. Gheber, V. Volterra, J. L. Hutchison, L. Margulis, R. Tenne, *J. Am. Chem. Soc.* **1994**, *116*, 1914.
[43] I. Dance, K. Fisher, *Metal Chalcogenide Cluster Chemistry*, in *Prog. in Inorg. Chem.*, Wiley, New York **1994**, *41*, pp. 637–802.
[44] J. A. Wilson, A. D. Yoffe, *Adv. Phys.* **1969**, *18*, 193.
[45] A. Aruchamy (Ed.), *Photoelechemistry and Photovoltaics of Layered Semiconductors*, Kluwer Academic Publishers, Dordrecht **1992**.
[46] Y. Feldman, E. Wasserman, D. J. Srolovitz, R. Tenne, *Science* **1995**, *267*, 222.
[47] D. J. Srolovitz, S. A. Safran, M. Homyonfer, R. Tenne, *Phys. Rev. Lett.* **1995**, *74*, 1779.
[48] D. Ugarta, *Nature* **1992**, 707.
[49] R. Tenne, L. Margulis, Y. Feldman, M. Homyonfer, *Proc. Symp. Fullerenes*, Materials Research Society, Fall Meeting, Boston, December **1994**, in press.
[50] a) Y. Saito, T. Yoshikawa, S. Bandow, M. Tomita, *Phys. Rev. B* **1993**, *48*, 1907. b) Y. Yosida, *Appl. Phys. Lett.* **1994**, *64*, 3048. c) M. Li, J. M. Cowley, *Ultramicroscopy* **1994**, *53*, 333.
[51] G. Frey, M. Homyonfer, Y. Feldman, R. Tenne, unpublished.
[52] a) S. Kolboe, C. H. Amberg, *Can. J. Chem.* **1966**, *44*, 2623. b) E. Furimsky, C. H. Amberg, *Can. J. Chem.* **1975**, *53*, 3567.
[53] a) Z. Shuxian, W. K. Hall, G. Ertl, H. Knözinger, *J. Catal.* **1986**, *100*, 167. b) C. B. Roxlo, M. Daage, A. F. Ruppert, R. R. Chianelli, *J. Catal.* **1986**, *100*, 176.
[54] Y. Feldman, M. Homyonfer, G. Hodes, R. Tenne, unpublished.
[55] a) P. D. Fleischauer, *ASLE Trans.* **1983**, *27*, 82. b) P. A. Bertrand, *J. Mater. Res.* **1989**, *4*, 180.
[56] a) A. R. Beal in *Physics and Chemistry of Materials with Layered Structures: Intercalated Layer Materials* (Ed: F. Levy), Redel, Dordrecht **1977**. b) H. Tributsch, *Structure Bonding* **1982**, *49*, 127.

Cage structures and nanotubes of NiCl$_2$

Nanoparticles of layered compounds form hollow cage structures of various types, including nanotubes[1]. Here we study nanoparticles of the layered compound NiCl$_2$, which can form cage structures, with different numbers of atomic layers, and nanotubes. Because Ni atoms in a layer are coupled ferromagnetically and the interlayer (IIc) coupling is antiferromagnetic, these structures may show different magnetic behaviour according to the parity of closed NiCl$_2$ layers in the nanoparticles.

NiCl$_2$ and other three-dimensional transition metal chloride compounds (MCl$_2$) crystallize in the layered CdCl$_2$ (space group $R\bar{3}m$) structure. Although within the layers there are strong mixed covalent–ionic bonds between the metal and halide atoms, the layers are held together by weak van der Waals forces. The magnetic moments of the Ni atoms are coupled ferromagnetically in the a–b plane. The moments of adjacent layers are oriented in opposite directions, producing an antiferromagnet with a Néel temperature of 52.3 K (ref. 2). NiCl$_2$ is a semiconductor with an energy bandgap of 2.4–2.8 eV (ref. 3).

Figure 1 shows transmission electron microscope images of typical closed cage structures, consisting of one closed atomic layer of NiCl$_2$ with a quasi-spherical (Fig. 1a) or a polyhedral shape (Fig. 1b). In another set of experiments, NiCl$_2$ cages with three and four layers were obtained.

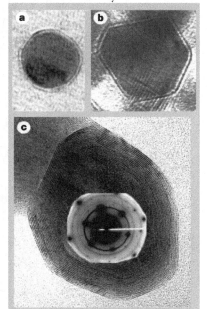

Figure 1 Transmission electron micrographs of NiCl$_2$ nanoparticles with a closed cage structure. **a,** Single-layer quasi-spherical particle (6 nm diameter); **b,** single-layer cage structure with polyhedral shape (10 nm diameter); **c,** many-layer cage structure. Superimposed on the hollow core is its hexagonal diffraction pattern. The distance between the NiCl$_2$ layers (fringes) is 0.58 nm ($c/3$). Nanoparticles were synthesized by heating hexahydrate in air at 150 °C until the powder changes from green to yellow. Dried powder is inserted into the reactor just outside the central hot zone, which is heated slowly to 400 °C under N$_2$ gas flow (50 cm^3 min^{-1}) for 30 min. The temperature of the hot zone is then raised to 960 °C. During heating the gas is slowly switched to Ar (50 cm^3 min^{-1}). After 30 min the boat is moved to the hot zone. When sufficient loose material has been collected (10–30 min), the temperature is reduced to 200 °C while the gas flow is maintained. The sample is desiccated on cooling to room temperature.

In both cases, the core of the nested cage structure does not appear to be hollow, although its exact nature is not known. A multiple-layer NiCl$_2$ hollow cage structure is shown in Fig. 1c. The hexagonal pattern of the selected area electron diffraction (SAED) of the core indicates that the nanoparticle is hollow, which is confirmed by high-resolution transmission electron microscopy (HREM). Energy-dispersive X-ray spectroscopy (EDS) using a probe less than 3 nm across revealed that the nano-particle consisted of nickel and chlorine in an atomic ratio of 1:2.

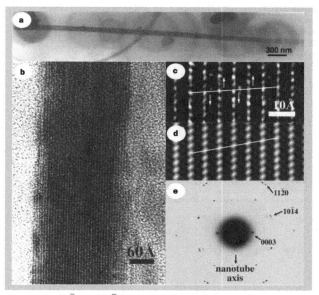

Figure 2 High-resolution transmission electron microscopy (HREM) analysis of NiCl$_2$ nanotube. **a,** Overall view of the nanotube (low magnification). **b,** Overall lattice image of the wall. The empty central channel is seen on its right. **c,** Higher, digitally filtered magnification of the sidewall region showing the image of eight $c/3$ planes of NiCl$_2$. Arrows indicate different displacements of the $c/3$ planes in the nanotube. **d,** HREM image simulation[4] of the same planes in bulk NiCl$_2$. **e,** SAED pattern of the nanotube. The {10$\bar{1}$1} and {10$\bar{1}$4} spots come from the planes perpendicular to the curved areas situated between the upper (lower) and side walls of the nanotube.

Figure 2 shows various HREM images of a NiCl$_2$ nanotube 6 µm long and with a cross-sectional diameter of 70 nm. Figure 2a shows an overall image of the nanotube, whereas Fig. 2b shows a lattice image of the tube wall on one side and the empty channel to its right. An HREM image of NiCl$_2$ layers of this tube wall and an image simulation[4] of bulk NiCl$_2$ in [01$\bar{1}$0] projection are shown in Fig. 2c and d, respectively. Note the difference in the lattice inclination stemming from the lateral displacements of the (0003) NiCl$_2$ layers (arrows) in the nanotube relative to NiCl$_2$ bulk.

The corresponding SAED, which has diffraction features typical of a cylindrical nanotube structure[5], is shown in Fig. 2e. The SAED can be interpreted as a composite diffraction pattern generated by different areas of the nanotube. The two strong {0003} spots correspond to diffraction from the planes of the side walls. The six {11$\bar{2}$0} spots are produced by the second prismatic planes perpendicular to the upper and lower walls of the nanotube. It can be deduced that the tube axis has [1$\bar{1}$00] direction, that is, it is at 30° with respect to the a axis. Some of the spots are doubled, indicating that the structure is chiral, with an angle of 1.8°. EDS indicates that this nanotube contains only nickel and chlorine. The nanotube was stable for a few days.

NiCl$_2$ is the first of a family of halogen compounds with a layered structure to be shown to form new nanostructures. By combining judicious principles of materials design and sophisticated measurement techniques, it is likely that new physical phenomena will be discovered in these nanomaterials.

Y. Rosenfeld Hacohen*, E. Grunbaum*, R. Tenne*, J. Sloan†, J. L. Hutchison†
**Department of Materials and Interfaces, Weizmann Institute, Rehovot 76100, Israel*
†*Department of Materials, The University of Oxford, South Parks Road, Oxford OX1 3PH, UK*
e-mail: cpreshef@weizmann.weizmann.ac.il

1. Tenne, R., Margulis, L., Genut, M. & Hodes, G. *Nature* **360**, 444–445 (1992).
2. Lindgard, P. A., Birgeneau, R. J., Als-Nielsen, J. & Guggenheim, H. J. *J. Phys. C* **8**, 1059–1069 (1975).
3. Ackerman, J., Fouassier, C., Holt, E. M. & Holt, S. L. *Inorg. Chem.* **11**, 3118–3122 (1972).
4. Stadelman, P. A. *Ultramicroscopy* **21**, 131–145 (1987).
5. Whittaker, E. J. W. *Acta Crystallogr.* **9**, 855–862 (1956).

Chapter 2
PREAMBLE TO THE SYNTHESIS

Around 1994 I realized that the time was ripe for a conceptual change whereby the mechanism of the synthetic reactions to produce the inorganic fullerene-like/nanotubes (IF/INT) would be studied in greater detail. This change reflected our belief that in order to study these phases systematically, we needed to synthesize them in larger amounts, and in a pure form. Around the same time, Yishai Feldman — a new immigrant to Israel, and an expert on X-ray diffraction measurements, joined my group. It became clear to us that while the chemical reactions involved are rather simple, the key element in our ability to divert the reaction from the thermodynamically stable bulk WS_2 product (platelets) to the IF/INT nanoparticles is the kinetic control. Obviously this clue is not unique to IF/INT, and in fact is a common element in the bottom-up synthesis of numerous other 0-D and 1-D nanostructures such as quantum dots; nanowires, carbon fullerenes and nanotubes.

In a series of very detailed papers (1.4, 2.1, 2.2) we were able to study the synthesis of a pure IF-MS_2 (M=Mo,W) nanoparticles phase, averting the formation of the bulk WS_2 platelets. Furthermore, we found that the key step in converting the WO_3 (MoO_3) nanoparticles to closed cages multiwall IF nanoparticles was a combination of fast surface reduction/sulfidation reactions. These reactions lead to the formation of a few closed sulfide layers engulfing a reduced oxide (MO_{3-x}) core, see **Fig. 2.1**. The few closed MS_2 layers passivate the nanoparticle surface and prevent necking and coalescence with neighboring nanoparticles. Subsequently, slow sulfur inwards diffusion and oxygen outwards diffusion lead to a gradual conversion of the oxide core into closed MS_2 layers. This reaction proceeded in a quasi-epitaxial fashion leading to (almost) perfectly crystalline closed-cage MS_2 nanoparticles. At this point Gary Hodes suggested using crude WO_3 powders as precursors for the IF/INT synthesis. This sounded initially like a bad idea, since we all knew that these powders are made of micron-size particles and hence would transform into platelets (bulk) WS_2 powder upon high temperature sulfidation. Nonetheless Yishai consulted with various suppliers and found that BDH powder was actually made of WO_3 nanoparticles. This was a true turning point in our efforts to scale up the synthesis of IF-WS_2 NP. In a matter of days Yishai was able to synthesize a few grams of the desirable IF nanoparticles. In particular it

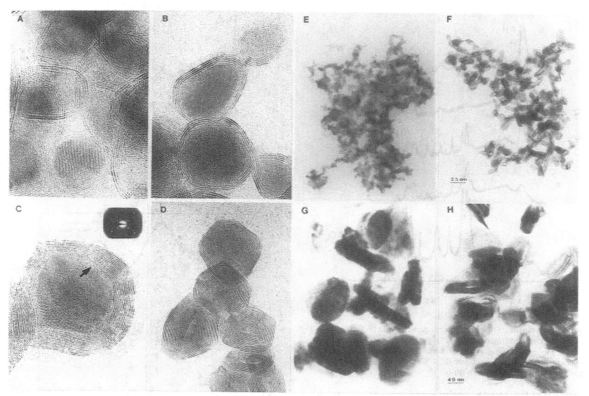

Figure 2. TEM micrographs showing the gradual transformation of molybdenum oxide nanoparticles into IF-MoS$_2$ (A−D) nested polyhedra. The electron diffraction pattern inset of (C) is consistent with (111) of MoO$_2$. (E) shows a typical assortment of tungsten oxide particles with <10 nm diameter which are transformed into IF-WS$_2$ particles of a similar size (Figure 2F). (G) and (H) show a similar transformation for tungsten oxide particles with a diameter in the range of 50−100 nm. Note that both oxide and IF phases contain very asymmetric particles. The interlayer spacing of 0.62 nm is clearly visible in (A)−(D).

After 2 min of annealing, a small WS$_2$ peak was observed (Figure 3B). Astonishingly, the entire nanoparticle core has been reduced to W$_{18}$O$_{49}$ at this early stage of the process. This fact can be understood assuming that hydrogen and water diffuse appreciably faster than sulfur diffuses to form sulfide. Figure 3C shows the state of the sample after 8 min of annealing, while Figure 3D displays the XRD patterns of the sample after 15 min of annealing time. The fully converted sample (120 min) is shown in Figure 3E. The shift of the (0002) peak of the IF-WS$_2$ phase (Figure 3D) indicates a lattice expansion of ca. 2% between two adjacent WS$_2$ slabs along the c-axis, which is attributed to the strain in the bent layers.[1] Furthermore, since the number of atoms in the layer increases with its radius, the layers cannot be fully commensurate. This discrepancy can be partially alleviated by lattice expansion along the c-axis. A similar study has been carried out for the conversion of MoO$_2$ into IF-MoS$_2$, the results being essentially the same as those shown in Figure 3.

Optical absorption spectra at different temperatures were measured for a sample consisting of IF-MoS$_2$ (shell)/MoO$_2$ (core) particles at different annealing times, including a pure IF-MoS$_2$ film (Figure 4A). A reference 2H-MoS$_2$ crystal was measured as well. Comparison of the spectra of IF-MoS$_2$ to that of the (2H) bulk phase at the same temperature reveals a red shift in the excitonic absorption of the former. The A, B, and C excitons appear at 667, 616.3, and 525 nm, respectively, in the IF phase at 175 K, while their energies in the 2H phase are 654.3, 593.5, and 490 nm.[23,24] After 3 min of annealing time of the oxide, the A and B excitons of the sulfide are already noticeable. The C exciton partially overlaps with the dominant absorption of MoO$_2$ at 500 nm.[25] The intensity of the oxide absorption decreased, while the exciton absorption increased with annealing time. After 90 min of annealing time, the exciton absorption peaks of the sulfide increased by a factor of 2.5 compared with the 3 min annealed sample. The oxide absorption disappeared completely. The transformation of WO$_{3-x}$ into IF-WS$_2$ (Figure 4B) was followed by using stirred alcoholic suspensions at room temperature. The oxide absorption peak of the composite oxide/sulfide (6 min annealing time) could be easily resolved in the difference spectrum and is substantially red shifted, compared to the literature value.[25] Remarkably, IF-MS$_2$ powder consisting of particles smaller than 10 nm formed a stable alcoholic colloid exhibiting a strong blue shift of the excitonic absorption, which could be possibly assigned to a quantum size effect.[24]

Being a surface sensitive technique, XPS could be ideally suited for investigating the sulfide/oxide superstructure. Accordingly, a few milligrams of IF-WS$_2$ powder, synthesized by the solid−gas reaction, was pressed onto an indium plate, which provided the support and electrical contact for the powder. A

(23) Wilson, J. A.; Yoffe, A. D. *Adv. Phys.* **1968**, *18*, 193.
(24) Frey, G. L.; Homyonfer, M.; Feldman, Y.; Tenne, R. To be published.
(25) Porter, V. R.; White, W. B.; Roy, R. *J. Solid State Chem.* **1969**, *1*, 359.

Synthesis of Inorganic Fullerene-like MS₂

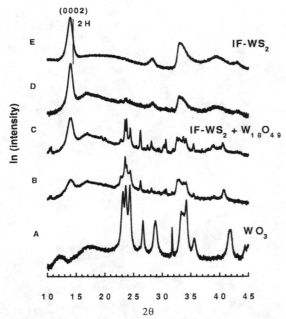

Figure 3. Transformation of WO₃ nanoparticles into IF-WS₂ followed by X-ray powder diffraction: (A) WO₃ precursor; (B) the same powder after 2 min of annealing; (C) after 8 min of annealing; (D) 15 min of annealing. Annealing conditions: 850 °C; rates of gas flow, forming gas, 130 cm³/min; H₂S, 2 cm³/min.

sequence of four samples consisting of composite nanoparticles of IF-WS₂ (shell)/tungsten oxide (core) at different annealing times at 850 °C were investigated: sample 1, 2 min; 2, 6 min, 3, 15 min; 4, 2 h. In addition, reference WS₂ crystal, WO₃ powder, and a clean indium specimen were measured. The fraction of oxide particle converted into sulfide was determined in two independent ways, which gave similar results. The concentration of the two compounds was determined first from the analysis of the total concentrations of the various elements, and second from the tungsten 4f–5p spectral region which was deconvoluted into sulfide (4f$_{7/2}$, 32.6; 4f$_{5/2}$, 34.75; 5p$_{3/2}$, 38.3 eV) and oxide (shifted to higher energies by 3.0 eV). Shirley background subtraction was used for spectral analysis.[26] While the indium contribution to the total signal was less than 1%, carbon made up to 18% of the total material. Yet, by comparing the molybdenum and sulfur line shapes with those of pure 2H-MoS₂, it was concluded that the carbon was not incorporated in the fullerene-like nanoparticles. The source of carbon contamination is likely to originate from adsorbed organic molecules having a mean thickness of 4–7 Å. Such an adsorbant signal does not seriously influence the analysis of its support, the fullerene-like material. Removal of the carbon contamination by ion sputtering was harmful to the IF structures and is not presented. Additional confirmation of that point came from local (TEM) electron energy loss measurements, which proved (due to its low sensitivity to surface contamination) that each nanoparticle consisted exclusively of Mo and S.

The spectra of 2H-WS₂ crystal and oxide powder, with line broadening permitted, were used as model line shapes for the deconvolution. The total error in this calculation was estimated at 1%. The high reliability of the deconvolution procedure was reaffirmed by using different pass energies. The results,

(26) *Practical Surface Analysis*, 2nd ed.; Briggs, D., Seah, M. P., Eds.; John Wiley & Sons: New York, 1990, Vol. 1, p 233.

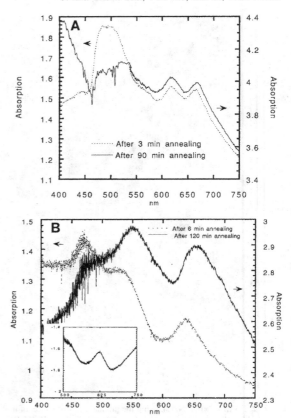

Figure 4. Optical absorption measurements of (A) a composite IF-MoS₂/MoO₂ sample obtained through the gas phase reaction and accrued on a quartz substrate after 3 min of annealing (dashed line) and after prolonged (90 min) annealing (solid line) and (B) a composite IF-WS₂/WO$_{3-x}$ sample after 6 min of annealing (dashed line) and after 120 min of annealing (solid line). The inset shows the difference spectra (oxide absorption peak). Annealing conditions are as in Figure 3.

Table 1. Conversion of Tungsten Oxide into IF-WS₂, Obtained from the Deconvolution of the W (4f) Peaks in a Series of XPS Spectra[a]

annealing time (min)	I_{ox} (%)	I_{sul} (%)	d/λ	k
2	38	62	0.6	2
6	11	89	1.5	5
15	1.5	98.5	3.3	11
120	<0.5	>99.5		

[a] Monochromatized Al Kα ($h\nu$ = 1486.6 eV) was used for the excitation. A monochromatized beam (0.7 eV resolution) and a base pressure of 10⁻⁹ Torr were used.

expressed as I_{sul} and I_{ox}, are summarized in Table 1. The relative intensities I_{sul}/I_{ox} can be converted into sulfide shell thickness (*d*) or number of dichalcogenide layers (*k*) by considering a model type polyhedron, for which the shell and core intensities are calculated face by face:

$$I_{sul}/I_{ox} = \sum W_n I_{n,sul} / \sum W_n I_{n,ox} \quad (1)$$

where W_n is the projection of the face area on the plane normal to the electron take-off direction (that is, parallel to the substrate plane). For simplicity we choose a regular polyhedron for the

Table 2. Kinetics of the Reduction of WO_3 Nanoparticles into Suboxides and its Conversion into IF-WS_2 Studied by XRD for Samples Prepared under Different Experimental Conditions

gas atmosphere; flow rate (cm^3/min)	av size of oxide particles(μm)	temp (°C)	reactn time (h)	product of reactn (XRD)	av particle size (μm) (TEM)
H_2S; 4.5 + H_2(5%)/N_2; 100	3–5	500	3	WO_3	not changed
		640	1	$WS_2 + W_{20}O_{58}$	0.1–3
		820	1	$WS_2 + W_{20}O_{58} + W_{18}O_{49}$	
		870	2	2H-$WS_2 + W_{18}O_{49}$	2–10
		970	1	2H-$WS_2 + WO_2 + W_{18}O_{49}$	
	~0.1	150	18	WO_3	not changed
		300	9	WS_2/WO_3	
		400	4	$WS_2/W_{20}O_{58}$	
		700	0.3	IF-$WS_2/W_{20}O_{58}$	
		820	0.3	IF-$WS_2/W_{18}O_{49}$	
		820	2	IF-WS_2	
		870	2	IF-WS_2 + 2H-$WS_2 + W_{18}O_{49}$	2–10
		970	1	2H-$WS_2 + W_{18}O_{49} + WO_2$	
H_2(5%)/N_2; 100	~0.1	560	2	WO_3	not changed
		570	1.5	$W_{20}O_{58}$	
		600	1	$W_{20}O_{58} + W + W_{18}O_{49}$	0.06–0.1
		650	1.5	$WO_2 + W$	0.03–0.1
		770	0.2	$W_{18}O_{49}$	0.1–1
		820	0.3	$W_{18}O_{49}$	1–2
		970	0.3	$WO_2 + W$	1–6
	3–5	630	1	WO_3	not changed
		650	1.5	$WO_3 + WO_2$	0.1–3
		750	0.2	$W_{20}O_{58}$	
		970	0.3	WO_2	1–6
H_2(1%)/N_2; 100	~0.1	600	1.5	WO_3	not changed

analysis. The two I_n terms in eq 1 are given by[27]

$$I_{n,\text{sul}} = \int_0^{d/\cos\theta_n} dz\, I_{0(\text{sul})} \exp(-z/\lambda \cos\theta_n) + \int_{D_n+d/\cos\theta_n}^{D_n+2d/\cos\theta_n} dz\, I_{0(\text{sul})} \exp(-z/\lambda \cos\theta_n)$$

$$I_{n,\text{ox}} = \int_{d/\cos\theta_n}^{D_n+d/\cos\theta_n} dz\, I_{0(\text{ox})} \exp(-z/\lambda \cos\theta_n) \quad (2)$$

The two terms in $I_{n,\text{sul}}$ stand for the contributions from the top and bottom shells of the nth segment (slope). I_0 is a factor proportional to the partial density of tungsten in the compound: $I_{0(\text{sul})} = I_0/3$; $I_{0(\text{ox})} = I_0/3.7$. D_n is the thickness of the oxide under the face of segment n. λ is the escape depth of the photoelectrons.

The calculated values for d/λ and k, which are given in Table 1, were obtained for a polyhedron having only six different slopes (segments): $\theta_n = 15°n$, $n = 0, ..., N - 1$. The number of slopes, $N = 6$, is a representative value. In fact the numerical result becomes only weakly dependent on N for $N > 3$. A similar model for spherical particles ($N \to \infty$) was recently published.[28] The value of λ was taken as ~2 nm[29] and that of the interlayer distance as ~0.62 nm.[23] The results of this analysis are in good agreement with the TEM observations; the main error stems from the variation in the number of MS_2 layers in the various nested polyhedra. XPS analysis of the MoO_2 into IF-MoS_2 conversion has been performed as well. Although the results of the analysis were qualitative in nature, the same trends as for the tungsten compounds have been observed.

The kinetics of the tungsten oxide reduction/sulfidization process was inferred from XRD measurements, and some typical results are presented in Table 2. From this table it is clear that, in addition to the experimental parameters, the size of the oxide particles plays an important role in the reduction/sulfidization reaction. For large (3–5 μm) particles, the reduction to the suboxide $W_{20}O_{58}$ and the subsequent sulfidization into WS_2 occur between 570 and 650 °C, whereas a powder consisting of small (0.1 μm) trioxide particles reacts at 400 °C. Furthermore, whereas the sulfidization of small particles occurs in a mode which leads to fullerene formation, the large particles form turbostratic-like WS_2 particles at temperatures lower than 600 °C, and 2H-WS_2 platelets upon sulfidization at elevated temperatures. In the absence of H_2S in the reactor, the reduction process is incomplete, and hence, it can be concluded that this gas serves as an auxiliary reducer of the oxide. Finally, the lower the concentration of hydrogen in the forming gas mixture, the slower was the reaction and the higher was the temperature required for the reaction. These considerations permit one to fine-tune the reaction conditions to the point that the IF-MS_2 phase can be synthesized with a very good reproducibility and yield.

Having the details of the reaction mechanism in mind, a modified reactor for the synthesis of macroscopic quantities of IF-WS_2 was constructed. The cross section of this reactor is illustrated schematically in Figure 5. To increase the amount of the (oxide) reactant and expose its entire surface to the gas, a bundle of tubes 7 mm in diameter each was placed inside the main tube (40 mm diameter) and the oxide powder was dispersed in them, very loosely. Typically, 1 g of IF-WS_2 could be obtained in a single batch, with a conversion yield of almost 100%.[30] This is remarkable in so far as the yield of production of carbon-nested fullerenes (and nanotubes) by the arc-discharge method is a few percent only.

In postulating a likely mechanism, we bear in mind the following: First, the reduction by H_2 is fast; the lower oxide is formed. Second, the formation of a thin sulfide skin controls the size of the fullerene. Since the size is thus fixed, the further transformation to sulfide must then take place internally.

A radial diffusion inward of the reactant (H_2S) and a similar diffusion outward of the product (H_2O) through the nanoparticle walls would appear unlikely in this case. Also, the large differences in time scales for the oxide reduction and the

(27) Reference 26, p 135.
(28) Sheng, E.; Sutherland, I. *Surf. Sci.* **1994**, *314*, 325.
(29) Reference 26, p 207.

(30) The major loss occurred from the evaporation of the oxide prior to the conversion with the first layer of the sulfide.

reduction is rather long and the sulfidization is accomplished before the reduction of the nanoparticles has started. At high temperatures, the first step of reduction from trioxide to suboxide is very quick and the sulfidization takes place only after reduction of the oxide nanoparticles. This situation further slows the rate of sulfur/oxide exchange, since much higher energies would be required for sulfur to break the stronger oxygen–metal bond of sub- or dioxide as compared with this bond in trioxide. The transition zone between the two regimes is discussed separately below.

The creation of oxygen vacancies on the surface of the oxide particles, by hydrogen, is most probably the first step in the entire (reduction–sulfidization) process (Figure 4A). Thence two competing events follow: the first one is vacancy annihilation by the shear process; the second event is sulfur occupation of the oxygen vacancy site. As noted above, sulfur occupation of the vacancy sites is more rapid than vacancy annihilation by a shear process at low temperatures (Figure 4B), while the reduction process prevails at high temperatures (Figure 4C).

The exponential dependence of the relaxation time of the shear process on temperature was discussed above. The relaxation time of the oxygen vacancy on the crystallite surface through sulfur occupation can be estimated from the kinetic theory of gases. The sulfur concentration defines an average distance between a vacancy on the particle surface and the nearest sulfur atom in the gas atmosphere. The kinetic energy of a particle in the gas atmosphere is equal to $k_B T$, and hence, the average velocity (V) of a sulfur atom is

$$V = (2k_B T/m)^{1/2} \quad (2)$$

where m is the mass of a sulfur atom. Thus, for a certain sulfur concentration, the relaxation time of oxygen vacancy by sulfur occupation is inversely proportional to the square root of the temperature.

4c. Synergistic Model. Figure 5 illustrates the schematic rates of reduction (A and B), sulfidization (C), and selenization (D) as a function of temperature. To plot this graphs, the functional dependence of the shear (eq 1) and sulfidization (eq 2) on temperature were calculated using arbitrary parameters, so that a single cross point between the two rates was obtained. Therefore the plots have qualitative significance only and they do not intend to represent a detailed calculation of the kinetic process. The sulfidization (selenization) reaction goes faster at low temperatures. Conversely, the reduction rate is much larger at high temperatures. The cross point of the reduction and sulfidization (selenization) curves shows the temperature range where the reduction rate is equal to the sulfidization (selenization) rate. It is suggested that instead of a competition between the two processes, a *synergy of the two processes exists* at this range of temperatures.

Figure 4 shows a schematic drawing of the shear (reduction) and sulfidization processes at different temperatures. The first step (step I) of vacancy formation by abstraction of one oxygen atom by a hydrogen atom is common to all temperatures. In the next step (II), any of three processes can take place; their relative rates vary with the temperature. At low temperatures, sulfur trapping is relatively faster than shear (Figure 4B). Conversely, at high temperatures, the shear process is faster (Figure 4C). At intermediate temperatures (600–850 °C), the rates of the two processes are equivalent (for a given experimental parameters this range is much narrower). During the shear of an octahedron, the oxygen atom, which hops from one corner to a neighboring vacancy, has to break its chemical bond to its nearest metal neighbor, and consequently, it is more prone to a chemical reaction or substitution at this instant of time. When the probability of oxygen leaving (reduction of oxide) equals the probability of the sulfur occupation of the oxygen vacancy site, the kinetic rate of the oxygen substitution by sulfur is the largest (Figure 4D). On the other hand, as noted above, the sulfur–oxygen exchange accelerates the autocatalytic reduction effect. Thus, a synergy between the autocatalytic reduction and oxygen–sulfur exchange prevails when the rates of the two processes are equivalent.

From the microscopic viewpoint, there is a significant difference between the low (high)-temperature processes and those at intermediate temperatures. In the low and high-temperature processes (Figure 4B,C), no oxygen atom is abstracted. In contrast to that, one oxygen atom is released when the synergy between the two processes (shear/sulfidization) occurs (Figure 4D). Moreover, this situation leads to the bonding of the captured sulfur atom to two neighboring M^V atoms, which is opposite to the situation at low temperatures. Subsequent abstraction of oxygen atoms and shear and concomitant sulfur atom capturing from the gas phase promotes the growth of the MS_2 surface layer encasing the oxide nanoparticle.

As shown above (section 3b), in the absence of hydrogen, sulfidization of the oxide leads to the formation of $2H-WS_2$. It is believed that the *synergy between the reduction and sulfidization processes provides the conditions for IF formation in a certain range of experimental parameters.* Following the formation of sulfide nuclei on the surface of an oxide nanoparticle, these two processes proceed simultaneously. The reduction process extends quickly inward by the fast rearrangements of the lattice and the formation of shear planes. This process leads to the "recrystallization" of the suboxide nanocrystal. On the other hand, oxygen/sulfur exchange cannot extend inward (into the particle core) due to steric hindrance of sulfur diffusion into the oxide core. The fast-growing sulfide layer replicates the morphology of the surface of the oxide nanoparticle. The anisotropy of the sulfide growth rate provides no time for a 3-D growth pattern of the nanocrystal and, thus, further promotes the formation of the curved sulfide 2-D (001) layer. It should be noted that the growth rate of the sulfide layer on the surface of the large oxide particles is not sufficient to wrap the entire surface, while reduction of the oxide takes place and they all break up into smaller crystallites (Table 2), which precludes IF formation from large oxide particles.

The formation of a *thin sulfide skin controls the size of the fullerene-like nanoparticle.* Once the size and shape are fixed, further transformation of the oxide into sulfide takes place internally. The growth of the next inner sulfide layers does not change the morphology of the outer layers (Figure 1). Therefore, sulfur and oxygen diffuse through available faults in the sulfide shell.[1]

The quasi-spiral inward growth process of the curved sulfide layers predetermines the hollow core of the fullerene-like nanoparticle. Every inner sulfide layer grows under a neighboring outer one. The radius of curvature of the innermost layers decreases as the process proceeds. The smallest radius of curvature of the sulfide layer, which corresponds to the minimum hollow core size, was found to be 1.4 nm.[13] However, most IFs have larger hollow cores.[13] The oxide core engulfed by the first closed sulfide shell is the precursor for the subsequent sulfide growth. The density of the crystalline metal oxide is higher than that of the metal sulfide. The presence of a hollow core in the IF particle suggests that the density of the primary oxide nanoparticles is much less than the formal density

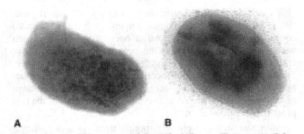

Figure 6. TEM images of polycrystalline (A) and onion-like (B) tungsten sulfide nanoparticles.

of the (monocrystal) oxide. Moreover, it is sometimes observed that the volume of the first shell of the sulfide layer (skin) can be appreciably larger than the volume of the starting oxide nanoparticle, leaving thus a free void between the sulfide skin and the oxide. Thus, the size of the hollow core of the IF is defined by the experimental conditions of this concrete nanoparticle.

As noted above,[1] the size of the precursor oxide particle defines the final dimensions of the IF particle. According to the synergy model (Figure 5), smaller oxide nanoparticles are wrapped by a sulfide layer at lower temperatures than bigger ones, which is confirmed by the experimental data (see Table 2). It is noticed that the size and shape of the IF particles follow closely those of the oxide precursor (see Figure 2 of ref 1).

Figure 6 shows TEM images of a polycrystalline (A) and fullerene-like (B) nanoparticles of WS_2. The polycrystalline nanoparticles were grown at 500 °C. This temperature is below the temperature corresponding to the *synergy effect*. The polycrystalline sulfide nanoparticle, like its oxide precursor nanoparticle, consists of very small randomly oriented nanocrystallites. Oppositely, the IF particle, which is formed via the *synergy effect* at much higher temperatures (800 °C), consists of a single domain and consequently has the long-range order of the curved atomic (metal disulfide) layers.

Synthesis of IF−WSe_2[2] shows also a good agreement with the present model. The cross point between the reduction and selenization processes (Figure 5) is observed at lower temperatures (typically 760 °C) compared with 840 °C for IF sulfide formation. Higher reaction temperatures yielded platelets (2H) of the respective selenides. Hence, the synergism between selenization and reduction, which is a prerequisite for the IF formation, occurs at lower temperatures than that of the respective sulfide. This tendency is confirmed by the available experimental data[2] and is further supported by preliminary data for IF of metal telluride's (630 °C). However, in this last case, it is unlikely that dislocation-free IF will be obtained at such low temperatures.

5. Conclusions

Direct evidence for the formation mechanism of the first fullerene-like layer of MX_2 (M = W, Mo; X = S, Se, Te) on the surface of oxide nanoparticle is demonstrated. This process is followed by a chalcogen/oxide exchange for which a direct proof is demonstrated on the same group of nanoparticles which were reacted step by step. It is shown that hydrogen is a necessary precursor for IF formation from metal oxide nanoparticles.

The synergy between the reaction of the chalcogen molecules coming from the gas phase and the reduction process of the oxide for IF−WX_2 (X = S, Se) synthesis is consistent with the available experimental data. Further work is necessary to provide a direct proof of this unique process.

Acknowledgment. We are grateful to Dr. J. Sloan of the University of Oxford for his useful suggestions regarding the SC process. This research was funded by the following grants: UK-Israel Science and Technology Foundation; NEDO (Japan) International research program; Israeli Ministry of Science (strategic research program on nanomaterials); Petroleum Research Foundation of the American Chemical Society; Minerva Foundation (Munich).

JA973205P

Growth Mechanism of MoS₂ Fullerene-like Nanoparticles by Gas-Phase Synthesis

A. Zak, Y. Feldman, V. Alperovich, R. Rosentsveig, and R. Tenne*

Contribution from the Department of Materials and Interfaces, Weizmann Institute, Rehovot 76100, Israel

Received June 19, 2000

Abstract: Inorganic fullerene-like (hollow onionlike) nanoparticles (IF) and nanotubes have attracted considerable interest in recent years, due to their unusual crystallographic morphology and their interesting physical properties. IF-MoS$_2$ and nanotubes were first synthesized by a gas-phase reaction from MoO$_3$ powder. This process consists of three steps: (1) evaporation of the MoO$_3$ powder as molecular clusters; (2) condensation of the oxide clusters to give MoO$_{3-x}$ nanosize particles; (3) sulfidization of the suboxide nanoparticles to generate IF nanoparticles. The evaporation of MoO$_3$ (step 1) and the IF particle formation from the oxide nanoparticles (step 3) have been investigated already, while the mechanism for the suboxide nanoparticles formation (step 2) has not been studied before and is reported here. According to the present model, a partial reduction of the trioxide molecular clusters (3–5 molecules) leads to the formation of MoO$_{3-x}$ nanoparticles (5–300-nm particles size)—the precursor for IF-MoS$_2$. A mathematical model, which takes into account the diffusion of the reactants into the reaction zone, the chemical reactions, and the boundary conditions obtained from the experiments, is established and solved. Based on the comprehensive understanding of the IF-MoS$_2$ growth mechanism from MoO$_3$ powder and the solution of the diffusion equations, a gas-phase reactor, which allowed reproducible preparation of a pure IF-MoS$_2$ powder (50 mg per batch) with controllable sizes, is demonstrated.

Introduction

Condensation of molecular clusters from the vapor phase is a conventional method for nanosize particle formation. In this method, a hot vapor is quenched and entrained by a flowing inert gas. Nanoclusters are obtained by an adiabatic expansion of the vapor leading to a cooling of the clusters inert gas vapor and its condensation.[1-3] Carbon fullerenes and nanotubes were obtained by a similar synthetic process.[4]

Fullerene-related nanoparticles are derived from materials with a layered structure. Each crystallographic layer of such a nanoparticle is closed into a quasi-sphere or a tubule structure. The weak van der Waals forces between the molecular layers and the strong intralayer covalent bonds is a necessary condition for the formation of these unusual nanoparticles. There are three main types of fullerene-related particles: fullerenes (C$_{60}$, C$_{70}$, etc.), nested-fullerene nanoparticles (onions), and nanotubes. Analogous fullerene-like nanoparticles were obtained from a number of inorganic materials with layered structure, which were designated as inorganic fullerene-like materials (IF).

Fullerenes are produced from carbon-rich vapor or plasma, which can be obtained using laser ablation, arc discharge, or resistive heating of graphite targets.[4-6] The synthesis of such structures implies curvature of very small atomic sheets and annihilation of the dangling bonds of the peripheral atoms. It was shown that fullerene-like nanoparticles of MoS$_2$ could be obtained using such methods as e-beam irradiation[7] and laser ablation[8] of regular MoS$_2$ powder. Short electrical pulses from the tip of a scanning tunneling microscope over a film consisting of amorphous MoS$_3$ nanoparticles lead to the formation of nanoparticles with a closed MoS$_2$ shell (IF), a few molecular layers thick, and an amorphous MoS$_3$ core.[9] Closed cages and nanotubes of NiCl$_2$ were recently observed upon heating NiCl$_2$ to 960 °C in a reducing atmosphere.[10] All the methods described above can be categorized as "physical" ones, since they do not exploit a chemical reaction for nanoparticle formation.

IF nanoparticles, including nanotubes, can be obtained also by "chemical" methods, in which case a chemical reaction is essential for the nanoparticles' growth. The first synthesis of IF-MS$_2$ (M = Mo,W) was based on the sulfidization of the respective amorphous MO$_3$ thin films in a reducing atmosphere at elevated temperatures (~850 °C).[11-13] Using molybdenum oxide powder instead of a thin-film precursor, IF-MoS$_2$, including MoS$_2$ nanotubes were reported.[14] This synthesis however, resulted in miniscule amounts of the nanoparticles and a limited size control. More recently, macroscopic quantities

---

(1) Flagen, R. C.; Lunden, M. M. *Mater. Sci. Eng. A* **1995**, *204*, 113.
(2) Deppert, K.; Nielsch, K.; Magnusson, M. H.; Kruis, F. E.; Fissan, H. *Nanostruct. Mater.* **1998**, *10*, 565.
(3) Kruis, F. E.; Fissan, H.; Peled, A. *J. Aerosol Sci.* **1998**, *29*, 511.
(4) Dresselhaus, M. S.; Dresselhaus, G.; Eklund, P. C. *Science of Fullerenes and Carbon Nanotubes*; Academic Press: Inc.: New York, 1996; pp 1–6, 110–116.
(5) Kroto, H. W.; Heath, J. R.; O'Brien, S. C.; Curl, R. F.; Smally, R. E. *Nature* **1985**, *318*, 162.
(6) Kratschmer, W.; Lamb, L. D.; Fostiropoulos, K.; Huffman, R. *Nature* **1990**, *347*, 354.

(7) Jose-Yacaman, M.; Lorez, H.; Santiago, P.; Galvan, D. H.; Garzon, I. L.; Reyes, A. *Appl. Phys. Lett.* **1996**, *69* (8), 1065.
(8) Parilla, P. A.; Dillon, A. C.; Jones, K. M.; Riker, G.; Schulz, D. L.; Ginley, D. S.; Heben, M. J. *Nature* **1999**, *397*, 114.
(9) Homyonfer, M.; Mastai, Y.; Hershfinkel, M.; Volterra, V.; Hutchison, J. L.; Tenne, R. *J. Am. Chem. Soc.* **1996**, *118*, 33, 7804.
(10) Rosenfeld-Hacohen, Y.; Grinbaum, E.; Sloan, J.; Hutchison, J. L.; Tenne, R. *Nature* **1998**, *395*, 336.
(11) Tenne, R.; Margulis, L.; Genut, M.; Hodes, G. *Nature* **1992**, *360*, 444.
(12) Margulis, L.; Salitra, G.; Tenne, R.; Talianker, M. *Nature* **1993**, *365*, 113.
(13) Hershfinkel, M.; Gheber, L. A.; Volterra, V.; Hutchison, J. L.; Margulis, L.; Tenne, R. *J. Am. Chem. Soc.* **1994**, *116*, 1914.
(14) Feldman, Y.; Wasserman, E.; Srolovitz, D. J.; Tenne, R. *Science* **1995**, *267*, 222.

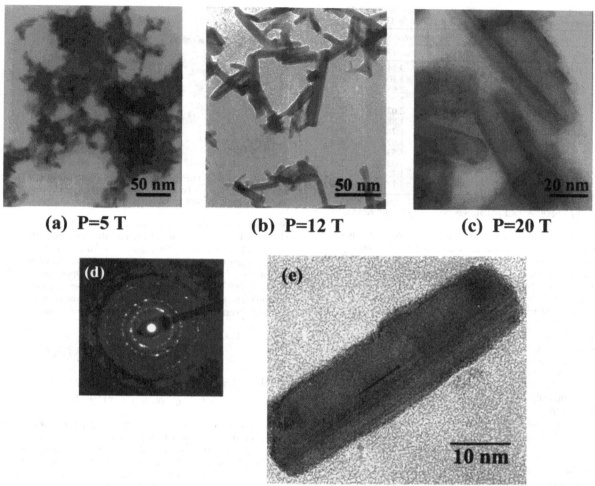

Figure 1. TEM micrographs and the corresponding ED patterns of tungsten oxide particles WO_{3-x} synthesized at different water vapor pressure: (a) $P_{H_2O} = 5$ Torr; (b) $P_{H_2O} = 12$ Torr; and (c) $P_{H_2O} = 20$ Torr. (d) ED of the WO_{3-x} particles synthesized at $P_{H_2O} = 12$ Torr and $P_{H_2O} = 20$ Torr. (e) Enlargement of one whisker produced at $P_{H_2O} = 12$ Torr. Defects along the whisker axis are indicated.

was maintained constant during the evaporation process (a few Torr). After a few minutes of evaporation, a blue powder condensed on the bell-jar walls. The accrued powder consisted of needlelike WO_{3-x} particles (ca. 50 nm in length and 15 nm in diameter) under a specific water vapor pressure (vide infra).

$NiCl_2$ or $CoCl_2$ (2×10^{-3} M) salts were dissolved in the water reservoir before each evaporation. The nanoparticles produced in the presence of the transition-metal salt appeared to be more crystalline than those obtained without the addition of a salt, as shown by ED.

II.2. Synthesis of the WS_2 Nanotubes Starting from the WO_{3-x} Nanoparticles. The synthesis of the WS_2 nanotubes starting from the needlelike WO_{3-x} particles was done in a reactor similar to the one used for the synthesis of IF-WS_2 particles.[13] The principle of the synthesis is based on a solid–gas reaction, where a small quantity (5 mg) of WO_{3-x} particles (solid) is heated to 840 °C under the flow of H_2/N_2 (forming gas) + H_2S gas mixture. To avoid cross-contamination between the different runs and minimize memory effects, which can be attributed to the decomposition of H_2S and deposition of sulfur on the cold walls of the reactor, flushing of the reactor (10 min) with N_2 gas flow was performed after each synthesis.

Samples were studied using a scanning electron microscope (SEM) Philips XL30-ESEM FEG instrument, a transmission electron microscope (TEM) Philips CM 120 (120 keV), and an X-ray diffraction instrument (Rigaku Rotaflex RU-200B) having a Cu Kα anode. The electron diffraction (ED) patterns were obtained on a high-resolution transmission electron microscope (HRTEM) JEM-4000EX operated at 400 kV. Ring patterns from TiCl were used as a calibration reference standard for the ED patterns. The accuracy of the d spacings is estimated at ±0.005 nm.

III. Results

III.1. Synthesis of Tungsten Oxide Needlelike Particles (Stage I). Three different values of water vapor pressure were selected: $P_{H_2O} = 5$, 12, and 20 Torr, the latter corresponding to the thermodynamic limit of the water vapor pressure at room temperature (22 °C). The texture of all the batches appeared to be more or less the same after a few minutes of evaporation. However, a variation in the color of the powder, which was collected on the walls of the bell-jar, was noticed. A color range, which goes from dark blue for $P_{H_2O} = 5$ Torr to light blue for $P_{H_2O} = 20$ Torr was observed.

Apparently, the water vapor pressure in the chamber influences the morphology and the stoichiometry of the nanoparticles obtained by evaporation. For a low value ($P_{H_2O} = 5$ Torr), the oxide nanoparticles did not have a well-defined morphology (Figure 1a). The ED pattern confirms that the powder is completely amorphous (not shown here). When the pressure was increased ($P_{H_2O} = 12$ Torr), the nanoparticles presented a

Growth of WS₂ Nanotubes Phases

Table 1. Comparative d Spacing Data between the Needlelike Precursors and the Tetragonal WO$_{2.9}$ Reported by Glemser[15a],[a]

oxide precursors		tetragonal WO$_{2.9}$[15a]		
I_{rel}	d_{hkl} (Å)	I_{rel}	d_{hkl} (Å)	hkl
100	3.752	100	3.74	110
20	3.206	20	3.10	101
80	2.640	80	2.65	200
30	2.184	30	2.20	201
		10	2.02	211
30	1.878	30	1.88	220
10	1.703	10	1.78	300
60	1.558	60	1.67	310
50	1.153	50	1.53	311
		10	1.33	222
		10	1.25	330
		10	1.17	322

[a] the d_{hkl} spacings were obtained from the ED ring pattern of the oxide particles. A TiCl pattern was used as a standard reference.

cylindrical shape and were crystalline. A typical batch is shown in Figure 1b, where the dimensions of the whiskers are typically around 50 nm in length and 15 nm in diameter. For the thermodynamic limit of the water pressure at room temperature (P_{H_2O} = 20 Torr), a growth in both directions (along the nanoparticle long axis and perpendicular to it) led to the formation of needlelike particles with a much smaller aspect ratio and steps perpendicular to the long axis. The whiskers are crystalline as could be evidenced from the ED pattern, which is similar to the one observed for the particles produced at 12 Torr (see Figure 1c).

The stoichiometry of the particles could not be easily assigned by XRD for several reasons. First, most of the samples were not sufficiently crystalline for generating well-defined peaks in the spectrum. Moreover, several nonstoichiometric tungsten oxide phases have been reported in the literature and all of them exhibit very similar patterns. Consequently, assigning the stoichiometry of the concerned phase accurately from the XRD data was rather difficult. The measurement by electron diffraction of a bundle of individual needlelike crystals was more informative in this case. The values of the d_{hkl} spacings were calculated for the crystalline whiskers synthesized at P_{H_2O} = 12 and 20 Torr. Both sets of whiskers can be interpreted as having an average substructure similar to that of the reported tetragonal phase W$_{20}$O$_{58}$ (WO$_{2.9}$) originally described by Glemser et al.[15a] (Table 1). The needles can be described according to a substructure of WO$_3$ interspersed with defects attributable to random crystallographic shear planes occurring either parallel to the needle axis or, alternatively, at some angle to the beam direction as the needles are viewed in the HRTEM. Further evidence of the randomness of the defects occurring in the needles is given by the prominent diffuse streaking that is often observed in ED patterns obtained from these needles[15b] (see also Figure 1d). An example of a needlelike particle containing random defects is shown in Figure 1e. It is noteworthy to underline that, whatever the pressure inside the chamber, the batches appeared to be homogeneous in their morphology, providing needlelike particles of relatively constant oxide stoichiometry for a given preparation.

Hereafter, a detailed study of the conditions required for the whisker's growth was undertaken. The role of the water in this process was examined first.

III.1.a. The Role of Water. To get an idea of the role of water in the oxidation of the tungsten filament, evaporations

(15) (a) Glemser, O.; Weidelt, J.; Freund, F. *Z. Anorg. Allg. Chem.* **1964**, *332*, 299. (b) Sloan, J.; Hutchison, J. L.; Tenne, R.; Feldman, Y.; Tsirlina, T.; Homyonfer, M. *J. Solid State Chem.* **1999**, *144*, 100.

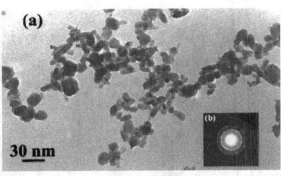

Figure 2. (a) TEM micrograph of tungsten oxide particles WO$_{3-x}$ synthesized at P_{O_2} = 6 Torr and (b) the corresponding ED pattern of the WO$_{3-x}$ particles synthesized at P_{O_2} = 6 Torr.

were performed with oxygen instead of water vapor in the chamber. Indeed, oxidation of the tungsten filament could be performed either with water vapor according to eq 1 or with pure oxygen (eq 2), both reactions being exothermic in the conditions of the present measurements (temperature of the filament, 1600 ± 20 °C; pressure in the chamber maintained at 12 Torr). The free energies of the reactions were calculated using the data given in ref 16 for STP (standard) conditions.

$$W (s) + 3H_2O (g) \rightarrow WO_3 (s) + 3H_2 (g)$$
$$(\Delta G_{(1873 K \text{ and } P=12 T)} = -21 \text{ kJ mol}^{-1})^{16} \quad (1)$$

$$W (s) + {^3/_2}O_2(g) \rightarrow WO_3(s)$$
$$(\Delta G_{(1873 K \text{ and } P=12 T)} = -150.5 \text{ kJ mol}^{-1})^{16} \quad (2)$$

To perform the evaporation with the same quantity of oxygen as for the one performed in the presence of water vapor, the oxygen pressure was maintained at P_{O_2} = 6 Torr compared to P_{H_2O} = 12 Torr (n_{O_2} = $^1/_2 n_{H_2O}$). The resultant particles were 100% spherical or faceted, typically 5 to 30 nm in diameter (see Figure 2). The color of the powder was light blue, which can be ascribed to a slight reduction of the powder by traces of water still present in the vacuum chamber. When the oxygen pressure was decreased, light blue phases of spherical or faceted nanoparticles were observed as well.

The absence of needlelike particles in the presence of oxygen in the chamber is indicative of the role played by hydrogen in generating an asymmetric growth of the nanoparticles (see eqs 1 and 2).

These findings allude to the fact that the needles growth consists of a two-step process occurring simultaneously on the hot filament surface. The first step is the oxidation of the tungsten filament, which leads to the formation of WO$_3$ particles. In the next step, reduction of these particles results in the formation of WO$_{3-x}$ needlelike particles (eq 3).

$$WO_3 (s) + H_2 (g) \rightarrow$$
$$WO_{3-x} (g) + xH_2O (g) + (1-x)H_2 (g) \quad (3)$$

It is important to note that the direct reaction between water vapor and the W filament is not the only plausible oxidation route. Indeed two pathways could be contemplated for the oxidation of W with water. The first one corresponds to the direct reaction of water molecules with W atoms (eq 1).

(16) *Handbook of Chemistry and Physics*, 69th ed.; Weast, R. C., Editor-in-Chief; CRC Press, Inc.: Boca-Raton, FL, 1988–1989; pp D50–93.

Table 2. Influence of an Auxiliary Gas on the Morphology of the Particles[a]

(a) Argon					
P_{H_2O} (Torr)	P_{Ar} (Torr)	P_{H_2O}/P_{Ar}	P_{H_2O}/P_{Tot}	P_{Tot} (Torr)	morphology of the particles
5	2.5	2	0.66	7.5	N-thin ($l \cong$ 50 nm and $D \cong$ 15 nm) + amorphous material
5	5	1	0.5	10	N-thin ($l \cong$ 50 nm and $D \cong$ 15 nm)
5	20	0.25	0.2	25	N-thin ($l \cong$ 90 nm and $D \cong$ 15 nm) + N-thick ($l \cong$ 90 nm and $D \cong$ 30 nm)
5	40	0.125	0.11	45	N-thin ($l \cong$ 110 nm and $D \cong$ 15 nm) + N-thick ($l \cong$ 130 nm and $D \cong$ 30 nm) + ⟨F−N⟩
12	12	1	0.5	24	N-thin ($l \cong$ 90 nm and $D \cong$ 15 nm) + N-thick ($l \cong$ 90 nm and $D \cong$ 30 nm)
12	48	0.25	0.2	60	⟨S−F⟩ ($D \cong$ 25 nm) + few N-thick ($l \cong$ 50−300 nm and $D \cong$ 35−180 nm)
(b) Hydrogen					
P_{H_2O} (Torr)	P_{H_2} (Torr)	P_{H_2O}/P_{H_2}	P_{H_2O}/P_{Tot}	P_{Tot} (Torr)	morphology of the particles
5	5	1	0.5	10	N-thin
5	20	0.25	0.2	25	no evaporation
12	1	12	$\cong$12	13	N-thin with more defects along the needle's axis
12	5	2.4	0.7	17	N-thin with more defects along the needle's axis

[a] N denotes particles with a needlelike morphology. l denotes length of the particles. D denotes diameter of the particles. ⟨F−N⟩ denotes particles with a morphology of between faceted and needlelike. ⟨S−F⟩ denotes particles between spherical and faceted.

Alternatively, partial water decomposition (see eq 4) leads to the oxidation of the hot tungsten filament by liberated oxygen.

$$H_2O\ (g) \rightarrow \tfrac{1}{2}O_2\ (g) + H_2\ (g)\ (\Delta G_{(1873\ K\ and\ P=12\ T)} = +33.9\ kJ\ mol^{-1})^{16}\quad (4)$$

Regardless of whether the direct or indirect mechanism is correct, H_2 is a resultant product of both reactions. It is therefore believed that hydrogen is involved in the production of needlelike particles as opposed to the spherical ones, which are obtained in the absence of hydrogen in the chamber.

III.1.b. Attempts to Increase the Size of the Needlelike Particles. In this series of experiments, the effect of different gases on the growth mode of the nanoparticles was examined. Since thermodynamics dictates that water vapor pressure cannot exceed 20 Torr at room temperature, addition of different gases (Ar or H_2) was attempted. Argon and hydrogen have a similar mean free path in the prevalent conditions ($\lambda_{Ar} = 4.73\ 10^{-5}$ m and $\lambda_{H_2} = 8.81 \times 10^{-5}$ m at 20 °C and 1 Torr);[16] however, hydrogen is directly implicated in the tungsten oxide whisker growth, while argon is chemically inert. The two sets of experiments were therefore used as a tool for underpinning the importance of the chemical nature of the gas in the process. Indeed, in addition to water (partial vapor pressure of P_{H_2O} = 5 and 12 Torr), argon was introduced first into the chamber in different partial pressures. Note that for P_{H_2O} = 12 Torr, oxide nanowhiskers are formed, while a partial pressure of 5 Torr of water leads to the formation of amorphous tungsten oxide nanoparticles.

The results obtained with the addition of argon in the chamber are summarized in Table 2a. It appears that needlelike particles are formed under 5 Torr of water vapor pressure complemented by 5 Torr of argon. This is rather surprising since the same partial pressure of water alone led to the formation of an amorphous material (see Figure 1a). Keeping the partial pressure of water constant, the argon pressure was elevated step by step, until a total pressure of 60 Torr was reached. The morphology of the whiskers begins to change already around 25 Torr, whence a second growth direction, perpendicular to the whisker's main axis, starts to evolve. Here, a stepwise growth mode is apparent, with new terraces added to the incipient whisker outer perimeter. This kind of "thick whisker" growth mode is similar to the results obtained for P_{H_2O} = 20 Torr (see Figure 1c). A mixture of thick whiskers and faceted particles could be observed above a total pressure of 25 Torr. Finally, when the total pressure had reached 60 Torr, essentially spherical or faceted particles (25 nm in diameter) and some remaining needlelike particles with steps (50−300 nm long and 35−180 nm thick) were present in the sample.

Accordingly, to have a pure phase of needlelike tungsten oxide particles, the total pressure has to be kept in the range 10−25 Torr, with at least 5 Torr of water vapor pressure.

The same procedure was repeated with an additional pressure of hydrogen instead of argon (Table 2b). As can be seen, the partial pressure of hydrogen also influences the morphology of the tungsten oxide particles. Indeed, for a total pressure of 25 Torr (P_{H_2O} = 5 Torr), no evaporation of the metal could be detected whatsoever, although for the same total pressure with argon instead of hydrogen, whiskers were formed (see Table 2a). This result emphasizes the fact that the nature of the gas has an important influence on the morphology of the particles. According to eq 1, addition of hydrogen into the vacuum chamber shifts the reaction to the left, thereby blocking the formation of WO_3 and its subsequent sublimation (at 1600 °C). When the ratio of the partial vapor pressures $P_{H_2O}/P_{H_2} \geq$ 0.5, the needlelike morphology is nevertheless conserved (see Table 2b). In this case, the reducing power of hydrogen is not strong enough to halt the formation of WO_3 nanoparticles, but is sufficient to promote the growth of the needlelike particles. Note also that, under these conditions (P_{H_2O} = 12 Torr and P_{H_2} = 1 or 5 Torr), a much larger density of random defects was observed in the oxide needles as compared to those produced by evaporation performed in the absence of hydrogen (Table 2b).

Accordingly, when a reactive gas such as hydrogen is introduced into the chamber, either directly (as H_2) or indirectly (as H_2O), both the reactive gas pressure (P_{gas}) and the total pressure in the chamber (P_{Tot}) have to be considered.

Unfortunately, no improvement in the aspect ratio and the length of the needles was observed in all the sets of experiments described in this section.

III.1.c. Attempts to Increase the Needlelike Particle Length via High-Temperature Reaction (Stage II). Since hydrogen was found to be indispensable for the growth of the needles, an alternative procedure for promoting their growth under more controllable conditions was pursued. The basic idea was to promote the uniaxial growth of the short tungsten suboxide needles obtained in stage I under very low hydrogen gas concentration. For that purpose, the needles were placed in a reactor operating at around 840 °C in a flow of (H_2/N_2) gas mixture where the concentration of hydrogen was progressively

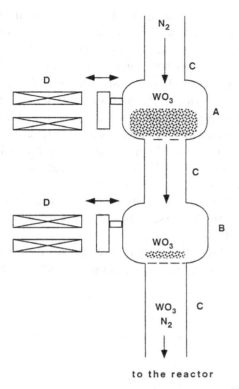

Fig. 2. Two stage vibro-electromagnetic feeding set-up.

4. The fluidized bed reactor

The feeding of oxide powder to the new reactor is carried-out by the same set-up (Fig. 2) used for the falling bed reactor. A generalized scheme of the fluidized bed reactor for the IF synthesis is shown in Fig. 3a. The temperature profile along the reactor is shown in Fig. 3b. The reactor consisted of a quartz tube (50 mm diameter), which was placed in a vertical three (heating) zones furnace. The oxide powder assisted by nitrogen gas flow (40 cc min^{-1}) fell inside a 9 mm diameter quartz tube (A, Fig. 3a) from the feeding system through the upper heated zone and down to the reaction zone. The main gas stream rate of H_2S and N_2–H_2 gases was about 350 cc min^{-1}. This stream entered the reactor through the bottom of the reactor and flowed through a middle tube (B, C, Fig. 3a) up. The upper part of this tube (B, Fig. 3a) is 40 mm in diameter, while the remainder of the tube (C, Fig. 3a) is 12 mm in diameter. Therefore the linear gas flow rate in the wide (upper) part of the tube is ca. 30 cm min^{-1}, which is about ten times slower than that in the narrow part of the tube. Nitrogen flow with oxide powder, which leapt out the tube A, interfused with a flow of H_2S and N_2–H_2 gases in the wide part of the tube B. The sulfidization of the oxide nanoparticles started at this moment, since the diffusion of the H_2S and N_2–H_2 into tube A was hindered by the fast nitrogen stream (100 cm min^{-1}), which was mixed with the oxide powder. Then the powder fell slowly through the wide part of tube (B, Fig. 3a) against a main gas stream down till it was lowered to the narrow part of the tube (C, Fig. 3a). High linear rate of the main gas stream in the narrow tube (C, Fig. 3a) offered fluidized bed conditions for the falling powder. The filter, which was placed inside the tube at the end of the constant temperature region at 830°C (Fig. 3b), collected a small part of the powder that dropped on the filter. Most of the powder fluidized in the space above the filter in a narrow but rather long tube (C, Fig. 3a). A small part of the powder was swept by the main stream and was collected on another filter, which was placed outside tube C at the end of the constant temperature region at 830°C (Fig. 3b) on the way of the gases exit. However, the largest amount of the material stayed in the fluidized bed region. In this way more than 50 g of closed and hollow IF-WS$_2$ powder could be synthesized per one batch (ca. 10 h). The latter reactor, notwithstanding a few necessary changes, can be considered as a mock-up for the industrial production of IF-WS$_2$.

5. Results and discussion

The main advantages of the new methods of IF-WS$_2$ synthesis compared with the previous method [5] are the following:
1. The falling bed and especially the fluidized bed setup resulted in IF-WS$_2$ having more perfect (spherical) shape (Fig. 4) than the product of the previous synthetic tools. This observation is attributed to the fact that the reaction takes place in the gas phase, where an isotropic environment for the reaction prevails.
2. The maximum size of IF particles, which can be obtained in these methods, are more than 0.5 μm (Figs. 4 and 5), compared with 0.2 μm in the previous reactor [5]. This means that much larger oxide nanoparticles could be converted into IF when they flow in the gas stream.
3. The vertical posture of the oven allows addition of the oxide powder into the chamber during the

reaction, continuously. The production rate of the falling bed setup is about of 1–2 g h^{-1} of a pure *IF* phase. No other phase is observed in the product.

4. The maximum production yield per batch is about 20 g of a pure *IF*-WS$_2$ phase by the falling bed reactor and is more than 50 g of a pure *IF*-WS$_2$ phase by the fluidized bed reactor. Moreover, the fluidized bed concept lends itself for scale-up and to production of appreciably larger amounts of a pure *IF*-WS$_2$ phase.

5. Pure *IF* phase with no extra contaminants are obtained in these processes. Therefore, expensive and time-consuming filtrations of purification processes are avoided.

As was mentioned above, the growth mechanism of *IF*-WS$_2$ was elucidated in quite a detail [5,6]. It was found that under certain conditions the simultaneous reduction and sulfidization of oxide nanoparticles lead to *IF*-WS$_2$ formation. However, if some parameters are changed and either of these two processes prevail, fullerene-like particles will not be obtained. In particular, a deviation of the hydrogen–sulfur concentration ratio from the specified value (1:1) at any point of the reaction chamber will most likely lead to 2H-WS$_2$ platelets formation there. Therefore, it is very important, especially at the first instant of the reaction, that a homogeneous mixture of the three gases would engulf each oxide nanoparticle.

Such kind of problems have been around also in the previous synthetic method [5]. Although oxide

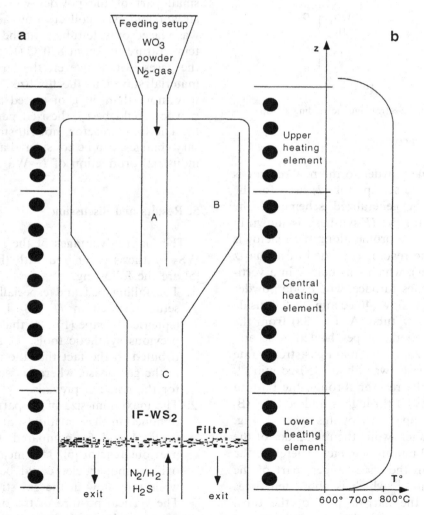

Fig. 3. (a) Schematic representation of the fluidized bed reactor; (b) temperature profile along furnace (z) axis.

powder was placed in a very loose form on the quartz substrate, sometimes rather dense lumps of the oxide nanoparticles could be found there. Higher diffusion of hydrogen compared with H$_2$S inside a lump lead often to a fast agglomeration of the oxide nanoparticles, which in turn resulted in 2H-WS$_2$ platelets. Using oxide powder having nanoparticles size smaller than 0.2 µm allowed us to avoid accidents of this kind, almost entirely. However, a decisive solution for this problem is found by using the falling bed and especially the fluidized bed reactors.

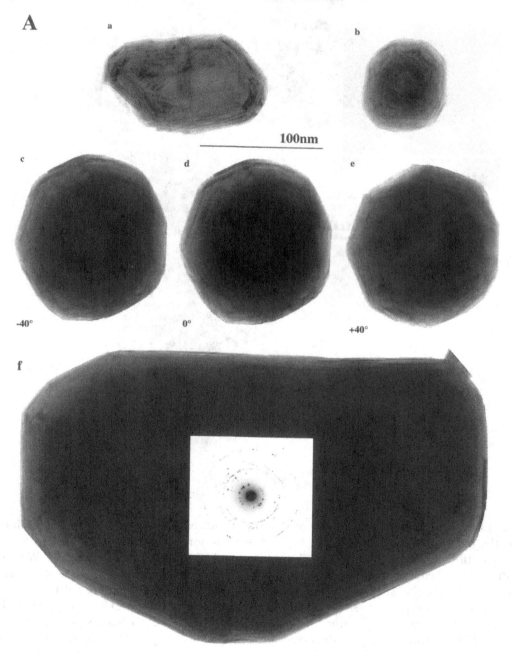

Fig. 4. TEM image of the typical *IF*-WS$_2$ nanoparticles: (a) the previous horizontal set-up; (b–f) the falling bed reactor; (c–e) three tilt projections of the same particle; (f) a large *IF*-WS$_2$ (400 nm) is shown with its electron diffraction. (B) High magnification of two *IF*-WS$_2$ nanoparticles (a, b of A part).

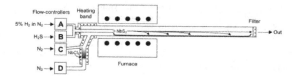

Fig. 1 Schematic drawing of the reactor used for the synthesis of the NbS$_2$ nanoparticles. NbCl$_5$ powder is placed on a quartz frit and heated to 200 °C. N$_2$ gas (flow-controller D) takes the vapor up. It is diluted with N$_2$ from flow-controller C. In the main reactor (furnace at 400 °C) the NbCl$_5$–N$_2$ gas mixture meets a mixture of forming gas (5% H$_2$, 95% N$_2$; flow-controller A) and H$_2$S (flow-controller B) coming out of a nozzle. There a chemical reaction occurs forming Nb$_{1+x}$S$_2$ and HCl. The fine Nb$_{1+x}$S$_2$ powder precipitates on the boat and the filter. The gases are taken out to washing bottles.

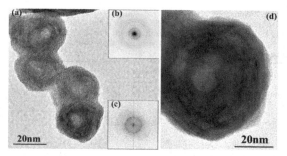

Fig. 2 TEM micrograph of a few NbS$_x$ nanoparticles with a non-perfect *IF* structure after the reaction. (a) Image of a group of a few *IF*-NbS$_2$ nanoparticles; (b) SAED of a group of such nanoparticles; (c) FFT of one of the nanoparticles; (d) large *IF*-NbS$_2$ nanoparticle obtained in the same reaction.

In fact, the value of x varied between $0 < x < 1$. Stable intermediates NbSCl$_3$ and NbS$_2$Cl$_2$, which decompose into NbS$_2$ upon heating, were identified.[16] Premature reaction of the NbCl$_5$ vapor and H$_2$S gas would generally lead to a product with a poorly defined composition. To allow the NbCl$_5$ vapor and H$_2$S gas to heat up before reaction and produce distinct chemical phases, the two gases were made to mix with each other close to the middle of the reactor (see Fig. 1).

The reaction products were collected on a filter inserted at the colder end (downstream) of the reactor and analyzed using transmission electron microscopy (TEM; 120 keV); X-ray energy dispersive analysis (EDS); and selected area electron diffraction (SAED). A scanning electron microscope (SEM) equipped with a EDS detector was also used for the analysis of the product.

Results

Table 1 summarizes the results of two out of many experiments, in which the synthesis of the NbS$_2$ nanoparticles according to Eqn. 1 was attempted. Fig. 2 shows TEM micrographs of a representative product from sample #1. Nanoparticles of close to spherical shape and closed nature (*IF*-NbS$_2$) are obtained. The nanoparticles are not fully detached from each other and they are also not perfectly crystalline containing numerous edge dislocations. Two kinds of *IF* nanoparticles are obtained in this synthesis. One family consists of relatively small nanoparticles (<30 nm), which is only partially crystallized (Fig. 2a–c). The other kind of nanoparticles (Fig. 2d) is larger (50 nm) and also more crystallized than the former kind. However, these nanoparticles contain numerous edge dislocations, and consequently although the NbS$_2$ layers are curved they are not fully closed. In fact more detailed analysis of the nanoparticles belonging to the first family in Fig. 2a reveals that they contain four distinct zones. The core of the nanoparticles does not seem to be hollow. Rather it consists of an amorphous (NbS$_x$) material. Further out there are a few folded but not fully closed NbS$_2$ layers. They appear to have started the crystallization process, but the short residence time in the furnace (<10 s) did not allow them to complete this process. On top of these layers, almost fully crystalline and closed NbS$_2$ layers can be discerned. These layers, however, intersperse with the NbS$_2$ layers of the adjacent nanoparticles making this whole agglomerate inseparable, even after a short ultrasonic treatment. Furthermore, a continuous top layer of amorphous material enfolds the nanoparticles and "glues" them together into an agglomerated superstructure. The nature of this amorphous coating does not seem to be different from the one in the nanoparticles' core. In general, the nanoparticles exhibit non-perfect crystallinity with numerous defects and some amorphous material interspersed with the crystalline NbS$_2$ layers. The composition of the nanoparticles cannot be easily verified. EDS analysis reveals that only Nb and S are present in the nanoparticles with an Nb : S ratio of approximately 1 : 1 for samples of this kind. The excess niobium in this sample is probably present in the amorphous core and the outer amorphous envelope of the nanoparticles. Note, however, that excess Nb intercalation between the NbS$_2$ layers[17] is not unlikely in this case. When the source temperature was reduced to 180 °C (sample #2 of Table 1), appreciably smaller (<20 nm) nanoparticles with quasi-spherical morphology and a higher degree of crystallinity were obtained.

As mentioned above NbS$_2$ may appear in two polytypes, hexagonal-2*H* (*P*6$_3$/*mmc*) and trigonal-3*R* (*R*3*m*). Due to the close similarity in the *d* values of the two structures it is rather difficult to assign the polytype of the particles from their electron diffraction ring pattern, however they seem to be closer to the trigonal 3*R* form. Analysis of the SAED pattern, obtained from a zone containing a few nanoparticles (Fig. 2b), shows that the average (00*l*) interlayer spacing is 6.15 Å. Fast Fourier transform (FFT) from images of individual particles, in the non-annealed sample (Fig. 2c), shows a rather diffuse ring, corresponding to interlayer spacing in the range 5.9–6.35 Å for the small (20–40 nm) particles and 6.2 Å for the larger (60–80 nm) ones.

In order to improve the crystallinity of the nanoparticles, the samples collected after the synthesis were annealed at 550 °C for various periods of times under H$_2$S and H$_2$ atmosphere. Table 2 shows the experimental parameters of the annealing of

Table 1 Experimental parameters used for the synthesis of the hollow NbS$_x$ nanospheres—step I. On the way through the furnace (15 to 20 seconds) NbCl$_5$ vapor reacts with H$_2$S and H$_2$ gas yielding amorphous but pre-structured nanoballs. The samples were collected from the filter at the end of the tube, which remained at room temperature throughout the reaction. The lower temperature at the heating band provided a lower concentration of NbCl$_5$ vapor. The nanoballs of sample #2 are smaller compared to sample #1

Sample #	T(NbCl$_5$ source)/°C	T(furnace)/°C	Flow-controllers			
			A: 5% H$_2$ (in N$_2$)/ml min^{-1}	B: H$_2$S/ ml min^{-1}	C: N$_2$ (diluting)/ ml min^{-1}	D: N$_2$ (carrier)/ ml min^{-1}
1	200	400	28	6	90	14
2	180	400	28	6	90	14

Table 2 Annealing of pre-structured nanoballs—step II. The nanoballs of sample #1 were annealed at 550 °C under a flow of a gas mixture of 5% H_2 (in N_2) and H_2S for various times. The crystalline structure became more and more ordered yielding hollow core *IF*-NbS_2 onion-like particles

Sample #	T(furnace)/°C	Flow-controllers A: 5% H_2 (N_2)/ ml min^{-1}	B: H_2S/ ml min^{-1}	Annealing time/ min
3	550	15	3	10
4	550	15	3	30
5	550	15	3	60
6	550	15	3	130
7	550	15	3	225

Table 3 Experimental parameters for the synthesis of Nb_2O_3 nanoflowers and sea urchin-like particles (samples #8 and #9), and also NbS_2 nanowhiskers (sample #10) with a modified apparatus

Sample #	T($NbCl_5$ source)/°C	T(furnace)/°C	Flow-controllers A: 5% H_2 (N_2)/ ml min^{-1}	B: H_2S/ ml min^{-1}
8	225	900	100	4.0
9	225	1050	100	4.0
10	225	400	25	5.0

sample #1. Fig. 3a shows a TEM image of a typical group of nanoparticles after 130 minutes of annealing. Fig. 3b shows an expanded view of one of the *IF*-NbS_2 nanoparticles. The annealed samples exhibit a few salient differences with respect to the non-annealed samples. First the amorphous halo around the nanoparticles has now partially disappeared and converted into fully crystalline NbS_2 layers. These layers further enfold the already existing closed NbS_2 layers, and also interconnect neighboring nanoparticles. Equally important is the fact that the amorphous layer in the center of the annealed nanoparticles has partially crystallized, too, and formed closed faceted NbS_2 layers, leaving an empty core in the center of the nanoparticle. EDS analysis of the samples annealed for 130 minutes shows that the Nb:S ratio approaches 1:2, *i.e.* $Nb_{1+x}S_2$. Samples annealed for less than 2 h presented an intermediate case between the fully annealed and the non-annealed samples, while samples annealed for longer periods of times exhibited a gradual coarsening, which stems from the close proximity of the agglomerated nanoparticles. Also, the density of edge dislocations is considerably reduced after the annealing and the nanoparticles are much more faceted. The inner hollow core of the nanoparticles exhibits distinct angles, most typically 90°. In several cases, the inner core seems to have formed an octahedron. Polyhedra of MoS_2 with octahedral structures have been suggested[2] and later reported.[18] They consist of six MoS_2 squares in the apices of the octahedra, and are believed to be the most stable form of MoS_2 nanoclusters. In contrast with MoS_2, in which the Mo atom is coordinated to the sulfur atoms in a trigonal prismatic arrangement, the Nb is coordinated to the S atoms in an octahedral arrangement. It is thought that the latter kind of coordination would prefer the octahedral polyhedra. However, more work is necessary to validate this concept.

Comparison of the electron diffraction patterns of the non-annealed particles and the annealed samples shows a shift in the (00*l*) layer spacing from 6.15 to 5.9 Å upon annealing (inset of Fig. 3a). FFT of images from a single nanoparticle of the annealed samples (inset of Fig. 3b) results in a pattern comprising sharp spots (layer spacing of 5.9 Å), including high order ones, which are indicative of a more ordered structure.

In a related synthetic approach amorphous and crystalline NbO_xS_y nanofibers were obtained. In this reactor the $NbCl_5$ vapor was mixed with forming gas (5% H_2; 95% N_2 at 100 ml min^{-1}) rather than with pure N_2 gas, before being swept to the main reactor. Table 3 shows the experimental conditions used for this series of experiments.

Figs. 4 and 5 show TEM micrographs of fibrillar nanostructures from samples #8 and #9. Fig. 6 shows SEM images of the same nanofibers (#8). The nanofibers seem to emanate from one center and grow in all directions thus forming 3-D hemispherical nanoflower patterns. The nanostructures were amorphous and did not produce any SAED pattern. EDS

Fig. 4 TEM micrographs of amorphous NbO_xS_y nanoflowers from sample #8 (synthesized at 900 °C). (a) Three particles stuck to each other by their intertwining thin needles; (b) longer and thicker needles obviously emanating from one center of nucleation.

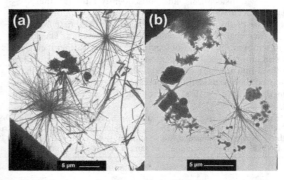

Fig. 5 TEM micrographs of amorphous NbO_xS_y nanoflowers from sample #9 (synthesized at 1050 °C). (a) The needles are much longer compared to Fig. 4; (b) the sample is not homogenous concerning the particle shape, nanoflowers with long needles are mixed with some sphere-like particles; particles with a bigger center exhibit shorter needles. This points to a root growth mechanism.

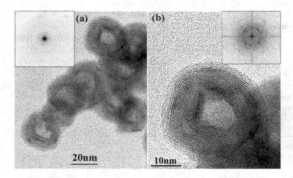

Fig. 3 TEM micrographs of the annealed *IF*-NbS_2 nanoparticles: (a) a group of such nanoparticles and their SAED in the inset; (b) higher magnification of one of the nanoparticles showing the closed nature of the layers and FFT of the nanoparticle in the inset.

Fig. 6 SEM micrographs of nanoflowers from sample #8, showing two typical morphologies: (a) sea urchin-like shapes; (b) star-like morphology, which seems to be a later level in the growth process, since the needles are longer and the sphere-like center is smaller.

analysis of the products was carried out. Short incipient nanofibers were composed mostly of Nb with little sulfur and oxygen. The composition of the developed (long) nanofibers varied between $NbO_{1.6}S_{0.1}$ and $NbO_{1.7}S_{0.2}$. This composition can be related on the lower (oxide) end to Nb_2O_3, which is a stable, but not so common oxide phase of niobium.[19] On the higher (O + S) end this composition can be associated with the much more common compound NbO_2. Similar patterns were previously obtained for SiO_x nanoflowers by firing SiC powder over a Co catalyst.[20] A similar growth pattern was observed also for carbon nanotubes grown from a central Co particle.[21] However, in the present work, no metallic catalyst is used *a priori*. Attempts to convert these nanofibers into NbS_2 nanotubes or crystallize them by annealing under different atmospheres and temperatures were unsuccessful. When the synthesis was carried out at lower temperatures (400 °C) and a slower flow rate of the forming gas (25 ml min^{-1}), crystalline NbS_2 nanofibers were obtained (see Fig. 7). Comparing the images from samples #8 and #9, one can see that at 900 °C (sample #8, Fig. 4) the needles have a length of less than 3 μm and the center of nucleation has a diameter of about 1 μm (black area in the TEM image). However at 1050 °C (sample #9, Fig. 5) much longer fibers can be observed (>10 μm) without a black area at the center; particles with shorter needles (top of Fig. 5b) similar to sample #8 are also found here. It looks like the material in the center serves as a supply for the needle growth (root growth). At higher temperatures the growth reaction is faster, thus—after this supply is consumed—the needles are longer.

Fig. 7 TEM micrograph of a crystalline NbS_2 nanowhisker from sample #10. The inset displays the SAED pattern of this nanowhisker, and reveals a layer distance of 5.99 Å in accordance with the distance of the lines measured from the image (5.9 Å). Investigation of the other reflections points to the 3R rather than the 2H phase, but a mixture of both cannot be excluded. Indices for the 3R phase are given in the inset.

Discussion

A rough estimate of the free energy change involved in Eqn. 1, using the data of ref. 22, shows that, while this reaction is almost balanced at room temperature, it becomes highly exothermic (< −150 KJ mol^{-1}) at 400 °C. In analogy to metathesis reactions,[23] which are highly exothermic at elevated temperatures, ignition of the reaction leads to a very rapid and sometimes explosive progression. The rapid reaction leads to a high degree of supersaturation of the vapor, and hence to a fast nucleation. The fast kinetics of the reaction together with the short residence time in the oven does not permit the crystallites to grow to a macroscopic size and consequently NbS_2 nanocrystallites with *IF* structure are formed.

It is very likely that amorphous NbS_x nanoparticles are formed in the vapor. The crystallization and the formation of the incipient *IF* structures occur before these nanoparticles agglomerate. Since the residence time of the nanoclusters within the hot zone of the reactor is not longer than a few seconds, the reaction of the amorphous NbS_x nanoparticles with H_2S and the crystallization of incipient *IF*-NbS_2 cannot be completed. Therefore, the nanostructures are heavily dislocated and the NbS_2 layers are not always fully closed. Further annealing of the product in H_2S and H_2 gases is necessary in order to produce a more perfectly crystalline phase. The annealing temperature cannot be too high, since otherwise coarsening of the agglomerated nanoparticles is unavoidable. Following 2–4 h annealing at 550 °C the nanoparticles become highly crystalline, but also faceted. Furthermore, the amorphous matter in the center of the nanoparticle has reacted with sulfur, forming new closed NbS_2 layers. The innermost NbS_2 layers, close to the now hollow core of the nanoparticles, become highly faceted with very sharp angles. In future work, synthesis in a fluidized bed reactor will be attempted. In this reactor the nanoparticles are kept aloft in the gas phase and do not interact with each other. This will allow annealing at higher temperatures without the danger of coarsening and loss of the *IF* structure.

A continuum theoretical analysis of the morphology of closed cage structures was undertaken.[24] This analysis suggests that closed cage structures of relatively large radius of curvature (>10 nm) and thin walls will bend uniformly and form quasi-spherical nanostructures. However, when the dimensions of the nanostructures go below a critical size the bending energy becomes excessively high, and point defects (grain boundaries) become energetically more favorable. Consequently a phase transition into a polyhedral structure is predicted in this case. This idea was validated by careful inspection of the growth of *IF*-WS_2 nanoparticles from tungsten oxide nanoparticles.[25] Here, the oxide to sulfide reaction proceeds from the outer surface of the oxide nanoparticle to the inner core. Hence the diameter of the first few closed WS_2 layers is large and they adopt a quasi-spherical shape. However, when the reaction proceeds, the oxide core is slowly consumed by the sulfide layers, which form closed WS_2 layers with an ever-smaller diameter. At a certain point faceted WS_2 polyhedral *IF* structures appear, which confirms the above theory. In the present case, the annealed samples are typified by having a thicker NbS_2 shell, *i.e.* a larger number of NbS_2 layers, than the non-annealed samples. Furthermore, the density of edge dislocations decreased after the annealing, making the faceting a more likely mechanism for the strain relief in the closed cage structures. These developments explain the transformation of the synthesized quasi-spherical NbS_2 nanoparticles into faceted polyhedra, after the annealing. Obviously, more work is needed to evaluate the most stable arrangements of NbS_2 polyhedral structures.

The formation of amorphous nanoflowers having a composition very close to Nb_2O_3 is quite unusual. In contrast to the previous works, which discussed the growth mechanism of

[3] Z. You, I. Balint, K.-I. Aika, *Appl. Catal. B* **2004**, *53*, 233–244.
[4] B. A. A. L. van Setten, C. G. M. Spitters, J. Bremmer, A. M. M. Mulders, M. Makkee, J. A. Molijn, *Appl. Catal. B* **2003**, *42*, 337–347.
[5] A. Simon in *Structure and Bonding*, Vol. 36 (Eds.: J. D. Dunitz, J. B. Goodenough, P. Hemmerich, J. A. Ibers, C. K. Jørgensen, J. B. Neilands, D. Reinem, R. G. P. Williams), Springer, New York, **1979**, pp. 81–127.
[6] K. R. Tsai, P. M. Harris, E. N. Lassettre, *J. Phys. Chem.* **1956**, *60*, 338–344.
[7] K. R. Tsai, P. M. Harris, E. N. Lassettre, *J. Phys. Chem.* **1956**, *60*, 345–347.
[8] a) A. Band, A. Albu-Yaron, T. Livne, H. Cohen, Y. Feldman, L. Shimon, R. Popovitz-Biro, V. Lyahovitskaya, R. Tenne, *J. Phys. Chem. B* **2004**, *108*, 12360–12367; b) S. Gemming, G. Seifert, C. Mühle, M. Jansen, A. Albu-Yaron, T. Arad, R. Tenne, *J. Solid State Chem.* **2005**, *178*, 1190–1196.
[9] C. N. R. Rao, M. Nath, *Dalton Trans.* **2003**, 1–25.
[10] R. Tenne, L. Margulis, M. Genut, G. Hodes, *Nature* **1992**, *360*, 444–446.
[11] L. Margulis, G. Salitra, R. Tenne, M. Talianker, *Nature* **1993**, *365*, 113–114.
[12] N. G. Chopra, J. Luyken, K. Cherry, V. H. Crespi, M. L. Cohen, S. G. Louie, A. Zettl, *Science* **1995**, *269*, 966–967.
[13] L. Rapoport, Yu. Bilik, Y. Feldman, M. Homyonfer, S. R. Cohen, R. Tenne, *Nature* **1997**, *387*, 791–793.
[14] Y. Q. Zhu, T. Sekine, K. S. Brigatti, S. Firth, R. Tenne, R. Rosentsveig, H. W. Kroto, D. R. M. Walton, *J. Am. Chem. Soc.* **2003**, *125*, 1329–1333.
[15] G. Brauer, *Handbuch der Präparativen Anorganischen Chemie*, Vol. 2, Enke, Stuttgart, **1978**, p. 938.
[16] A. Zak, Y. Feldman, V. Lyakhovitskaya, G. Leitus, R. Popovitz-Biro, E. Wachtel, H. Cohen, S. Reich, R. Tenne, *J. Am. Chem. Soc.* **2002**, *124*, 4747–4758.
[17] G. Seifert, H. Terrones, M. Terrones, M. Jungnickel, T. Frauenheim, *Phys. Rev. Lett.* **2000**, *85*, 146–149.
[18] I. Cabria, J. W. Mintmire, *Europhys. Lett.* **2004**, *65*, 82–88.
[19] P. O. Jeitschko, A. Simon, R. Ramlau, Hj. Mattausch, *Z. Anorg. Allg. Chem.* **1997**, *623*, 1447–1454.

Synthesis of fullerene-like MoS$_2$ nanoparticles and their tribological behavior†

R. Rosentsveig,[a] A. Margolin,[b] A. Gorodnev,[b] R. Popovitz-Biro,[c] Y. Feldman,[c] L. Rapoport,[d] Y. Novema,[e] G. Naveh[a] and R. Tenne[a]

Received 24th November 2008, Accepted 3rd February 2009
First published as an Advance Article on the web 6th March 2009
DOI: 10.1039/b820927h

Further understanding of the growth mechanism and the detailed structure of fullerene-like MoS$_2$ (IF-MoS$_2$) nanoparticles was achieved by using a new kind of reactor. The annealed nanoparticles consist of >30 closed layers and their average diameter is 50–80 nm although a small (<5%) fraction of larger IF nanoparticles was discernible. The majority of the nanoparticles are found to have an oval (pitta-bread or flying-saucer) shape rather than being quasi-spherical. The (002) peak of the powder diffraction pattern reveals only a small (0.3%) shift to lower angles as compared to the bulk (2H) phase. This observation suggests that the structure of the nanoparticles produced in the present reactor is more relaxed as compared to the previously synthesized IF-MoS$_2$ powder, which exhibited up to 2% shift. The present reactor also permitted scaling up of the production of the IF-MoS$_2$ to more than 0.6 g/batch. Impregnation of such nanoparticles in metallic coatings is shown to endow these surfaces with excellent tribological behavior, which suggests numerous applications.

Introduction

The discovery of inorganic nanotubes and inorganic fullerene-like nanoparticles (denoted INT and IF, respectively) of WS$_2$ in 1992[1] and subsequently that of MoS$_2$ in 1993[2] and 1995[3] marked the start of a new field in inorganic synthesis of nanomaterials and their applications. This field has been reviewed by a number of authors in recent years.[4–7] Initially, much of the research effort was focused on the need to elucidate the growth mechanism of IF-MS$_2$ (M = Mo,W) from the respective metal oxide precursors, which formed the basis for further study of their properties. Indeed, the progress in scaling up the synthesis of IF-WS$_2$ nanoparticles permitted their study as superior solid lubricants and their recent commercialization by "ApNano Materials" (www.apnano.com). In one of the early mechanistic studies,[8] a pure phase of IF-MoS$_2$ nanoparticles could be formed by heating MoO$_3$ powder and reacting the ensuing vapor with H$_2$S gas at >800 °C in a reducing atmosphere. This work allowed for some control over the diameter of the nanoparticles, but the amount of the synthesized material was typically 5 mg and below, which did not permit a systematic tribological study.

In the meantime, various groups have reported different strategies towards the synthesis of IF and INT of MoS$_2$. Early on, Yacaman and co-workers used electron beam irradiation of MoS$_2$ crystallites to obtain IF-MoS$_2$ nanoparticles.[9] Thereafter, Remskar, Mihailovic and co-workers have developed a chemical vapor transport technique permitting them to grow bundles of iodine-doped single wall MoS$_2$ nanotubes.[10] MoS$_2$ nanotubes stuffed with IF-MoS$_2$ nanoparticles in their hollow core (the so-called Mama-tubes) were obtained by sulfidizing Mo$_6$S$_2$I$_8$ nanowires at 930 °C in a reducing atmosphere.[11] MoS$_2$ nanotubes were also synthesized by direct reaction of (NH$_4$)$_2$MoS$_4$ and hydrogen.[12] Alternatively, MoS$_2$ nanotubes were prepared by heating MoS$_2$ powder to 1300 °C in the presence of a Mo foil and H$_2$S.[13] Refluxing (NH$_4$)$_2$Mo$_2$S$_{12}$ in acetone solution led to the formation of spherical MoS$_{5.6}$ precursor, which was converted to IF-MoS$_2$ nanoparticles upon further annealing to 800 °C.[14] Hollow imperfect spherical MoS$_2$ nanoparticles were also obtained by ultrasonic irradiation of a slurry containing Mo(CO)$_6$, sulfur, and nanometer silica spheres in isodurene under Ar gas flow. Subsequently, the silica core was removed with HF solution and final annealing at 450 °C in the presence of H$_2$S gas in a reducing atmosphere.[15] The nanoparticles were found to be very good catalysts for the mineralization of thiophene vapors at temperature >325 °C. Fast thermolysis of ammonium molybdate was shown to lead to a mixture of Mo oxides with spherical and rod-like morphologies. The oxide mixture was converted to IF- and INT-MoS$_2$ by H$_2$S reduction at 800 °C.[16] In another study by the same group Mo carbonyl was made to react with sulfur in a reducing atmosphere at 800 °C leading to IF-MoS$_2$ nanoparticles.[17] Particularly interesting were the experiments in which amorphous MoS$_3$ (WS$_3$) nanoparticles were obtained by the molecular organic chemical vapor deposition technique.[18] Upon annealing these nanoparticles *in situ* in the TEM up to 800 °C, they were shown to spontaneously convert into quasi-spherical IF-MoS$_2$ (IF-WS$_2$) nanoparticles. These experiments clearly indicate that the IF-MS$_2$ nanoparticles are (meta)stable on the border line between the amorphous MS$_3$ (M = Mo,W) phase and the bulk 2H-MS$_2$ phase. In another set of experiments IF-MoS$_2$ nanoparticles were synthesized at

---

[a]*Department of Materials and Interfaces, Weizmann Institute, Rehovot, 76100, Israel*
[b]*NanoMaterials Ltd., Weizmann Science Park, P.O. Box 4088, Nes Ziona, 74140, Israel*
[c]*Chemical Research Support Unit, Weizmann Institute, Rehovot, 76100, Israel*
[d]*Holon Institute of Technology, Department of Science, P.O. Box 305, Holon, 58102, Israel*
[e]*Scientific Glass Blowing Unit, Weizmann Institute, Rehovot, 76100, Israel*
† This paper is part of a *Journal of Materials Chemistry* issue in celebration of the 75th birthday of C. N. R. Rao.

relatively low temperatures (600 °C) by reacting ammonium molybdate with sulfur in hydrogen atmosphere.[19] Hydrothermal synthesis (180 °C) of INT-MoS$_2$ was also described,[20] but the quality of the product was a far cry from the nanotubes synthesized at elevated temperatures. Synthesis of MoS$_2$ nanoparticles with "fullerene-like" structure via low temperature (150 °C) solvothermal reaction was reported.[21] The synthesized nanoparticles were fully characterized by various techniques. Indeed, relatively large microspheres (200–3000 nm) were obtained by this method. The MoS$_2$ powders were mixed in oils at 0.25–0.5 mass% concentrations and were found to reduce friction and wear more than twofold. Arc discharge was also used to produce films of IF-MoS$_2$ nanoparticles on various substrates.[22] These films exhibited excellent tribological behavior. Later on[23] this group improved the IF-MoS$_2$ nanoparticles synthesis and achieved much better size and shape control, by immersing the arc-discharge set-up in water. The water mediates the energetic arc reaction and allows the nanoparticles to grow under better controlled conditions and therefore much more uniformly. Near-monodisperse microscopic nanotubes, however with fair crystallinity only, were prepared by thermal decomposition of ammonium thiomolybdate molecular precursors, within the confined voids of a porous aluminium oxide membrane template and annealing to 450 °C.[24] Laser ablation was used quite extensively to obtain MoS$_2$ nano-octahedra which are probably the smallest closed hollow nanostructures of that compound, i.e. the "true inorganic fullerenes".[25,26] In an important theoretical paper, Seifert et al.[27] have calculated the band-structure of MoS$_2$ nanotubes and showed that it varies with the role-up vector (n,m) of the nanotube. While the arm-chair (n,n) nanotubes were shown to exhibit an indirect transition, the zig-zag (n,0) nanotubes were shown to be direct bandgap semiconductors. No experimental verification of this important conclusion is available at this time.

The growth mechanism of IF-MoS$_2$ from MoO$_3$ powder was studied in greater detail in ref. 28, where the synthesis was separated into four consecutive steps: 1. evaporation of the MoO$_3$ powder at 700–750 °C; 2. partial reduction of the MoO$_3$ vapor into incipient MoO$_{3-x}$ nanoparticles at 780 °C; 3. fast sulfidization of the first few layers of the oxide nanoparticles in the vapor phase. In the 4th and final stage completion of the sulfidization reaction occurs when the nanoparticles land on the collecting filter. A slow diffusion-controlled reaction at >840 °C leads to the conversion of the entire MoO$_2$ oxide core into closed MoS$_2$ layers, which grow inwards layer by layer in a quasi-epitaxial fashion. Obviously, the incipient MoO$_{3-x}$ nanoparticles covered by a few MoS$_2$ closed layers, which are produced in situ, provide the template onto which the oxide core of the nanoparticle is gradually converted into IF-MoS$_2$ nanoparticles. Thus, the size of the IF nanoparticles reflects the size of the incipient sub-oxide nanoparticles. This study indicated that controlling the rates of MoO$_3$ powder evaporation and the subsequent reduction of the oxide vapor into MoO$_{3-x}$ nanoparticles determine the size of the IF-MoS$_2$ nanoparticles. Since this diffusion-limited step in the 4th step is the rate limiting process of the entire reaction, it is rather important to understand the structure of this oxide and the rate of oxide to sulfide conversion. It was shown before that under too aggressive reduction conditions the oxide core is reduced to metallic (Mo), which is encapsulated by a few closed layers of MoS$_2$.[29] In this case the sulfidization reaction ceases after two or three MoS$_2$ layers have been completed and the compact metallic core remains unchanged. The compact and dense metallic core (10.2 g/cm^3) does not permit the sulfidization reaction to proceed any further and the formation of the less dense (5 g/cm^3) MoS$_2$ layers ceases. Therefore, it became obvious that the reduction conditions must allow, on the one hand, nucleating MoO$_{3-x}$ nanoparticles from the MoO$_3$ vapor. But at the same time excessive reduction to dense Mo core prevents the sulfidization reaction to go to its end. Thus mild reducing conditions are essential for the formation of a relatively open structure oxide core, which allows for the completion of the conversion of the nanoparticle into a hollow IF-MoS$_2$ structure. This subtle reduction balance is further discussed in the present work. It is found that the oxide core consists of relatively closed packed MoO$_2$, which can not be easily converted to close MoS$_2$ layers. Thus, in contrast to the case of IF-WS$_2$, a prolonged (20 hr) annealing period is needed to bring the reaction to completion. On the other hand if the temperature is raised beyond say 1000 °C, the nanoparticles are converted to platelets of 2H-MoS$_2$ thereby leaving a relatively narrow temperature window for the formation of hollow IF-MoS$_2$ nanoparticles.

The improved understanding of the growth mechanism was further exploited in the present experiments, where a scaled-up version of the "fluidized-bed" reactor (FBR)[28] was constructed permitting synthesis of about 600 mg/batch of high quality IF-MoS$_2$ nanoparticles with limited size control. The synthesized nanoparticles were further annealed to convert the oxide core into closed layers of molybdenum disulfide. This sequence of steps was followed by various techniques. Many of the nanoparticles exhibited oval (pitta-bread or flying saucer) shape. The IF-MoS$_2$ powder was further used for fabricating self-lubricating metallic coatings. These coatings exhibit low friction and wear loss behavior which is comparable to their IF-WS$_2$-based predecessor coatings, but with somewhat reduced chemical stability, especially at elevated temperatures (>300 °C). Their lighter weight (5.0 versus 7.5 g/cm^3 for WS$_2$) and higher Young's modulus, combined with the low friction and wear properties can nevertheless be advantageous for a variety of applications.

Experimental

Synthesis of the nanoparticles

The fundamental reaction parameters were not that different from those employed in ref. 28. However, the reactor design was altogether different, permitting synthesis of appreciably larger amounts of the IF-MoS$_2$ nanoparticles in the pure form. Furthermore, the shape of the nanoparticles (oval vs. quasi-spherical in ref. 28), the number of closed MoS$_2$ layers (>25) and the very small empty core in the center are characteristic of the present powder. These observations also stand in quite a contrast with the quasi-spherical IF-WS$_2$ where the hollow core in the center occupies 30% of the nanoparticle volume and above.[30] Furthermore, under the favorable conditions described below, no macroscopic platelets could be observed in the reaction product, although some larger (>200 nm in diameter) IF particles were inadvertently obtained. Note, however, that during the first 15 min when the reactor is inserted into the furnace and heats up, the product of the reaction is not pure IF. It can nevertheless be

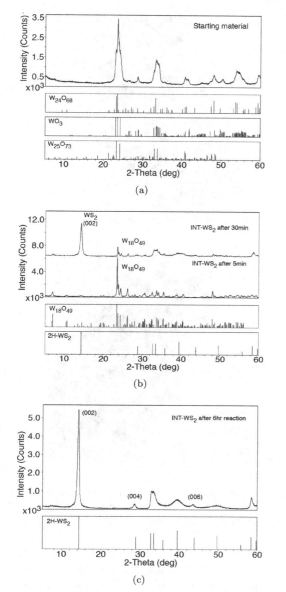

Fig. 3. (a) XRD pattern of the starting material. (b) XRD pattern of the product after 5 and 30 min reaction time. Note that after 5 min reaction time, the pattern coincides with the $W_{18}O_{49}$ phase only, while after 30 min reaction time, the sulfide phase coexist with the oxide one. (c) XRD pattern of the final product (after 6 h reaction time). A pure WS_2 phase is seen. Note also that the (002) peak at 14.13° is shifted to lower angles (higher spacing along the c-axis) as compared to the bulk material (14.32°).

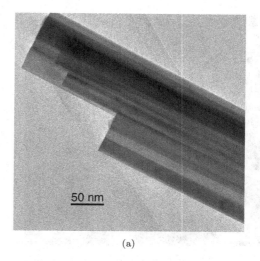

Fig. 4. TEM pictures (a) and (b) of variety of tungsten oxide nanowhiskers after 1 min reaction period. No sulfide layers are discernible at this stage of the reaction.

show the TEM images of a variety of oxide nanowhiskers after 1 min reaction time. No evidence for sulfide was found in this product. After 5 min reaction time (not shown), some of the nanowhiskers became coated with a few sulfide layers on their outer surface. During this period of time the sulfide layers did not have time to fully encapsulate the entire surface and they engulf a small part of the nanowhisker, only.

Figure 5(a) shows a HRTEM image and electron diffraction of a nanotube after 30 min reaction time. The ordered shear planes are clearly revealed in the core of the nanotube. The electron diffraction is a superposition of the diffraction from the oxide and the sulfide. The 004 and 010 (pointed with arrow) reflections belong to the $W_{18}O_{49}$ oxide phase which coincide with XRD results. Figure 5(b) shows the growth of the sulfide layers in its intermediate stage. Here, the few sulfide layers engulfing the oxide core are discernible.

(a)

(b)

Fig. 5. TEM micrographs of a WS$_2$ nanotube with highly ordered tungsten oxide core after 30 min reaction time. The electron diffraction (a) is a superposition of the diffraction from the oxide and the sulfide. The 004 and 010 (pointed with arrow) reflections belong to the W$_{18}$O$_{49}$ oxide phase. The closed outer tungsten sulfide layers in their intermediate stage (b) represent the incipient nanotube growth.

TEM images of typical nanotubes after 6 h reaction time is shown in Fig. 6. The nanotube exhibits highly crystalline order and contains many (> 20) layers with quite narrow inner empty core. Such nanotubes are expected to reveal very good mechanical properties.

4. Discussion

The growth mode of the nanotubes appears to first involve the formation of long, tens of microns,

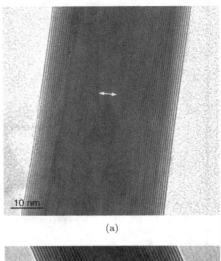

(a)

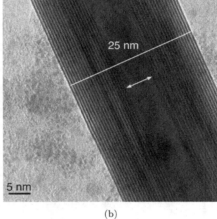

(b)

Fig. 6. TEM images of the WS$_2$ nanotubes after 6 h reaction time. The diameters of the tubes are (a) ~ 45 nm and (b) 25 nm. The double-sided arrows indicate the small width of the hollow core of the nanotubes.

suboxide (W$_{18}$O$_{49}$) nanowhiskers. This oxide serves as a template for the subsequent sulfidization reaction, producing eventually hollow WS$_2$ nanotubes. It was shown that the nanotubes, prepared in the large modified FBR, grow and sulfidize separately, according to the second type of the growth mechanism, described in the introduction. However, in this case, both reactions take place in the same reaction volume, and the temporal separation between them occurs by self-control led process and due to the significant difference between the reaction rates of reduction/growth and sulfidization. The impetuous growth of nanowhiskers through oxide reduction, sublimation, and condensation occurs at the first stage of the reaction, which excludes the sulfidization process.

The rapid growth of the oxide nanowhiskers is likely to go through a volatile oxide intermediate. The starting material consists of a mixture of oxides phases with the most reduced phase being $W_{24}O_{68}(WO_{2.83})$. After 10 min reaction time and beyond, the majority of the nanotubes contain the oxide phase $W_{18}O_{49}(WO_{2.72})$ in their core. These observations suggest that the oxide to tungsten ratio in the volatile phase has a composition of between these two extremes. It appears that the growth of nanowhiskers ceased after the entire oxide was reduced to the nonvolatile $W_{18}O_{49}(WO_{2.72})$ phase. In this work, the volatile oxide phase is not necessarily associated with water as suggested before.[6,15] Moreover, $W_{18}O_{49}$ is known as the most stable reduced suboxide phase. Thus, once formed, further reduction of the nanowhiskers is inhibited and the sulfidization reaction is started by a self-controlled led process.

The sulfidization proceeds similar to the reaction of IF[6,9,16,21]: the oxide gradually converts to sulfide starting from the external closed cylindrical layers inwards by a diffusion control reaction. The empty core obtained at the end of the reaction is the result of the difference between the oxide- and sulfide-specific densities. The formation of the closed external sulfide layers on the surface of the oxide nanowhisker at the initial stages of the sulfidization fixes the tube diameter. However, the density of the oxide (7.15 g/cm^3) is lower than that of the sulfide (7.5 g/cm^3) resulting in the formation of an empty core. Since the densities difference between the oxide and sulfide is rather small, the obtained empty core is relatively narrow. Contrary to this result, relatively wide empty core and a few sulfide layers were only obtained when the growth of INT occurs according to the first type of mechanism described in the introduction.[6,7]

The complete conversion of the oxide into sulfide was achieved after 6 h reaction time under the present parameters. The relatively slow oxide to sulfide conversion can be attributed to a number of factors. First, much of the nanowhiskers possess a relatively large diameter (> 100 nm) and the rate of diffusion of sulfur towards the inner oxide core of the nanowhisker is rather slow. Another factor is based on topological effects. The density of defects in the sulfide layer determines the rate of sulfur diffusion towards the core of the nanoparticle, i.e., the rate of oxide to sulfide conversion. This density is much higher in general in IF nanoparticles, which fold along two axes, as compared to nanotubes which fold along one axis only. It is also important to notice that the (002) and (004) peaks of the nanotubes are shifted to lower angles, i.e., the interlayer spacing increases by about 2% as compared to the bulk powder, due likely to the built-in strain in the nanotube lattice.[6,7,16]

It is believed that the sulfidization of the INT/2H mixture, which appear at the early stages of the reaction, affects the nanoparticles which do not participate in whiskers growth. These nanoparticles have plane or spherical shape and are undergone fast sulfidization.[16,17]

5. Conclusions

Reproducible mass production (50–100 g/batch) of a pure WS_2 nanotube phase together with larger amount (400–500 g/batch) of mixed nanotubes/platelets (50:50%) has been demonstrated using a fluidized-bed reactor. The nanotubes exhibit relatively wide size distribution (diameter of 20–180 nm and length of 1–150 microns). The nanotubes seem to grow through a volatile suboxide intermediate. This phase is likely to be in the composition range between $WO_{2.72}$ and $WO_{2.83}$. It is believed that careful control of the H_2 feeding and the reduction process could increase the length and the amount of the prepared oxide nanowhiskers and consequently the WS_2 nanotubes.

The oxide nanowhiskers first grow consisting of the stable $WO_{2.72}(W_{18}O_{49})$ phase. In the next step, the nanowhiskers serve as a template for the formation of the closed sulfide layers. The two reactions, growth/reduction and sulfidization, occur in the same reaction volume but are separated in time due to the reaction kinetics difference and exchange from one to the other by a self-controlled process.

Various new polymer nanocomposites are currently being studied with the IF-WS_2 nanoparticles as additives.[22,23] The availability of INT-WS_2 in large amounts, their high crystalline quality, and outstanding mechanical properties[19,20] pave the way for numerous similar studies, which may bring about various new applications for the nanotubes. For example, they could be used to strengthen various polymer nanocomposites, being more effective due to their high aspect ratio.

Acknowledgments

We are grateful to Drs. Y. Feldman, R. Popovitz-Biro and A. Albu-Yaron for their help with the analysis of the nanotubes. R. Tenne holds the Drake

and nanoscrolls was promoted by the relaxation of misfit stress between the adjacent SnS_2 and SnS layers. Partial decomposition of the SnS_2 precursor to a more sulfur-deficient SnS was stimulated by Bi and Sb_2S_3, which was manifested in the exfoliation of layers and scrolling. The yield of the nanotube synthesis was drastically enhanced by the addition of minute amounts of Sb_2S_3. Constant Bi and Sb profiles were observed along the tubules. Pure SnS_2 nanotubes were also formed, probably as a result of the annealing of preformed ordered-superstructure nanotubes. The presence of the two main structures was confirmed by HRTEM and Raman spectroscopy. Quantum mechanical calculations of the elastic energy of the SnS_2 and SnS nanotubes were carried out as a function of their diameter and chirality. It was shown that the structure of the ordered-superstructure nanotube is dictated by the low folding energy of the zigzag SnS (0,n) tube, which was indicated by electron diffraction analysis. Successful scaling up of the syntheses of SnS_2/SnS ordered-superstructure and pure SnS_2 tubes was accomplished with horizontal and subsequently vertical flow reactors. Possible applications for the present nanotubules can range from batteries[10] and photocatalysts[19] to toxic gas sensors.[20]

Experimental Section

Quartz ampoules were filled with SnS_2 (Alpha Aesar 99.5%) and Bi (Fluka 99.999%) powders (5:1 molar ratio). Minuscule amounts of Sb_2S_3 (Cerac/Pure 99.999%) powder was also used in several experiments. The ampoules were sealed in a vacuum of 2×10^{-5} torr and inserted into a horizontal two-zone reactor furnace. The performed high-temperature annealing procedure involved two main steps: First the whole ampoule was heated to 780°C for 2 h. Next, it was subjected to a temperature gradient of 780–250°C for 1.5 h, and was then cooled to room temperature. The product accumulated in the cold zone of the ampoule. Carbon/collodion-coated Cu TEM grids and SEM stubs based on Si/Al substrates were prepared by dripping several droplets from the product solution.

The resulting samples were examined by transmission electron microscopy (TEM; Philips CM120 operating at 120 kV, equipped with an energy-dispersive X-ray spectroscopy (EDS) detector (EDAX-Phoenix Microanalyzer) for chemical analysis), and by high-resolution TEM (FEI Technai F30-UT, with a field-emission gun operating at 300 kV). Scanning transmission electron microscopy (STEM; FEI Tecnai F20 operating at 200 kV equipped with high-angle annular dark field (HAADF) detector and (EDS) detector (EDAX-Phoenix Microanalyzer)), and also scanning electron microscopy (SEM; Zeiss Ultra model V55 and LEO model Supra 55VP equipped with EDS detector (Oxford model INCA) and backscattering electron (BSE) detector) were utilized. For Raman spectroscopy, a frequency-doubled Nd:YAG laser emitting $\lambda = 532.252$ nm light and a LabRAM HR 800 spectrometer (HORIBA, Jobin Yvon) with a 1800 mm grating and a spectral resolution of about 1 cm^{-1} were used. Individual nanotubes were placed on Si wafers (for SEM analysis) and also on carbon- and Si_3N_4-membrane-coated TEM grids (see the Supporting Information for the sample preparation).

The density-functional-based tight-binding method (DFTB) with periodic boundary conditions[21,22] was used for the quantum chemical calculations. In this approximation the LCAO ansatz (linear combination of atomic orbitals) is used for the molecular orbitals and a minimal set of atomic valence orbitals obtained from self-consistent density functional theory (DFT) calculations of the isolated atoms is taken as a basis. The repulsion is described by parameterized short-range two-body potentials.

Received: June 30, 2011
Published online: October 28, 2011

Keywords: chalcogenides · nanotubes · ordered superstructures · tin

[1] B. Alperson M. Homyonfer, R. Tenne, *J. Electroanal. Chem.* **1999**, *473*, 186–191.
[2] S. Y. Hong, R. Popovitz-Biro, Y. Prior, R. Tenne, *J. Am. Chem. Soc.* **2003**, *125*, 10470–10474.
[3] A. Yella, E. Mugnaioli, M. Panthofer, H. A. Therese, U. Kolb, W. Tremel, *Angew. Chem.* **2009**, *121*, 6546–6551; *Angew. Chem. Int. Ed.* **2009**, *48*, 6426–6430.
[4] J. Rouxel, A. Meerschaut, G. A. Wiegers, *J. Alloys Compd.* **1995**, *229*, 144–157.
[5] D. Bernaerts, S. Amelinckx, G. Van Tendeloo, J. van Landuyt, *J. Cryst. Growth* **1997**, *172*, 433–439.
[6] G. Falini, E. Foresti, M. Gazzano, A. F. Gultieri, M. Leoni, I. G. Lesci, N. Roveri, *Chem. Eur. J.* **2004**, *10*, 3043–3049.
[7] R. C. Sharma, Y. A. Chang, *Bull. Alloy Phase Diagrams* **1986**, *7*, 269–273.
[8] S. Del Buccia, J. C. Jumas, M. Maurin, *Acta Crystallogr. Sect. B* **1981**, *37*, 1903–1905.
[9] R. M. Hazen, L. W. Finger, *Am. Mineral.* **1978**, *63*, 289–292.
[10] S. Licht, G. Hodes, R. Tenne, J. Manassen, *Nature* **1987**, *326*, 863–864.
[11] M. Hangyo, K. Kisoda, T. Nishio, S. Nakashima, T. Terashima, N. Kojima, *Phys. Rev. B* **1994**, *50*, 12033–12043.
[12] Y. Shoji, T. Matsui, T. Nagasaki, M. Kurata, T. Inoue, *Int. J. Thermophys.* **2000**, *21*, 585–591.
[13] D. G. Mead, J. C. Irwin, *Solid State Commun.* **1976**, *20*, 885–887.
[14] A. J. Smith, P. E. Meek, Y. A. Liang, *J. Phys. C* **1977**, *10*, 1321–1333.
[15] A. Cingolani, M. Lugara, G. Scamarcio, *Nuovo Cimento Soc. Ital. Fis. D* **1988**, *10*, 519–528.
[16] H. R. Chandrasekhar, R. G. Humphreys, U. Zwick, M. Cardona, *Phys. Rev. B* **1977**, *15*, 2177–2183.
[17] P. M. Nikolic, L. Miljkovic, P. Mihajlovic, B. Lavrencic, *J. Phys. C* **1977**, *10*, L289–L292.
[18] K. Trentelmann, *J. Raman Spectrosc.* **2009**, *40*, 585–589.
[19] Y. C. Zhang, Z. N. Du, S. Y. Li, M. Zhang, *Appl. Catal. B* **2010**, *95*, 153–159.
[20] H. Chang, E. In, K. J. Kong, J. O. Lee, Y. Choi, B. H. Ryu, *J. Phys. Chem. B* **2005**, *109*, 30–32.
[21] D. Porezag, T. Frauenheim, T. Kohler, G. Seifert, R. Kaschner, *Phys. Rev. B* **1995**, *51*, 12947–12957.
[22] G. Seifert, D. Porezag, T. Frauenheim, *Int. J. Quantum Chem.* **1996**, *58*, 185–192.

Nanomaterials

Controlled Doping of MS$_2$ (M = W, Mo) Nanotubes and Fullerene-like Nanoparticles**

*Lena Yadgarov, Rita Rosentsveig, Gregory Leitus, Ana Albu-Yaron, Alexey Moshkovich, Vladislav Perfilyev, Relja Vasic, Anatoly I. Frenkel, Andrey N. Enyashin, Gotthard Seifert, Lev Rapoport, and Reshef Tenne**

Doping of semiconductor nanocrystals and nanowires with minute amounts of foreign atoms plays a major role in controlling their electrical, optical, and magnetic properties.[1] In the case of carbon nanotubes, subsequent doping with oxygen and potassium leads to a p-type and n-type behavior, respectively.[1a–c] In another work, VO$_x$ nanotubes were transformed from spin-frustrated semiconductors to ferromagnets by doping with either electrons or holes.[2]

Calculations indicated that n- and p-type doping of multiwall MoS$_2$ nanotubes (INT) could be accomplished by substituting minute amounts of the Mo lattice atoms with either Nb[3] (p-type) and Re[4] (n-type), respectively. Substituting (<0.1 at%) molybdenum by rhenium atoms[5] and sulfur by halogen atoms[6] was shown to produce n-type conductivity in MoS$_2$ crystals.

To synthesize rhenium-doped nanoparticles (NP) and nanotubes both in situ and subsequent doping methods were used. Figure 1a shows the quartz reactor used for in situ synthesis of rhenium doped MoS$_2$ NP with fullerene-like structure (Re:IF-MoS$_2$). The formal Re concentration was varied from 0.02 to 0.7 at%. The precursor Re$_x$Mo$_{1-x}$O$_3$ (x < 0.01) powder was prepared in a specially designed auxiliary reactor (see Supporting Information). Evaporation of this powder takes place in area 1 at 770°C (Figure 1a). The oxide vapor reacts with hydrogen gas in area 2 (Figure 1a) at 800°C which leads to a partial reduction of the vapor and its condensation into Re-doped MoO$_{3-y}$ nanoparticles. The resulting NP react with H$_2$/H$_2$S gas in area 3 at 810–820°C to produce reduced oxide nanoparticles engulfed with a few closed layers of Re:MoS$_2$, which protect it against ripening into bulk 2H-MoS$_2$.[7] To complete this oxide to sulfide conversion a long (25–35 h) annealing process at 870°C in the presence of H$_2$S and forming gas (H$_2$ 10 wt%; N$_2$) was performed. At the end of this diffusion-controlled process a powder of Re-doped MoS$_2$ NP with a fullerene-like (IF) structure (Re:IF-MoS$_2$) was obtained.

In addition, doping of IF-WS$_2$ NP and INT-WS$_2$ was subsequently carried out by heating the pre-prepared IF/INT in an evacuated quartz ampoule also containing ReO$_3$, or ReCl$_3$ and iodine. In the case of ReCl$_3$, both the rhenium and the chlorine atoms (substitution to sulfur atoms) served as n-type dopants.

Typical high-resolution scanning electron microscopy (HRSEM) and transmission electron microscopy (HRTEM) micrographs of the Re-doped fullerene-like NP are shown in Figure 1b. The Re:IF-MoS$_2$ consists of about 30 closed (concentric) MoS$_2$ layers. No impurity, such as oxides, or platelets (2H) of MoS$_2$ could be found in the product powder. The line profile and the Fourier analyses (FFT) (inset of Figure 1b) show an interlayer spacing of 0.627 nm (doped). Furthermore, the layers seem to be evenly folded and closed with very few defects and cusps, demonstrating the Re-doped NP to be quite perfectly crystalline. HRTEM did not reveal any structural changes even for the samples with high Re concentration (0.71 at%). However, owing to its quasi-spherical shape and size, this analysis cannot completely rule-out the presence of a small amount of the ReS$_2$ phase in the IF NP. Figure 1c shows a typical TEM image of Re(Cl) (post synthesis) doped multiwall WS$_2$ nanotube. There is no

[*] L. Yadgarov, Dr. R. Rosentsveig, Dr. A. Albu-Yaron, Prof. R. Tenne
Department of Materials and Interfaces, Weizmann Institute
Rehovot 76100 (Israel)
E-mail: reshef.tenne@weizmann.ac.il

Dr. G. Leitus
Department of Chemical Research Support, Weizmann Institute
Rehovot 76100 (Israel)

Dr. A. Moshkovich, Dr. V. Perfilyev, Prof. L. Rapoport
Department of Science, Holon Institute of Technology
P.O. Box 305, 52 Golomb St., Holon 58102 (Israel)

Dr. R. Vasic, Prof. A. I. Frenkel
Department of Physics, Yeshiva University
245 Lexington Ave, New York, NY 10016 (USA)

Dr. A. N. Enyashin
Institute of Solid State Chemistry UB RAS
Pervomayskaya Str., 91, 620990 Ekaterinburg (Russia)

Prof. Dr. G. Seifert
Physikalische Chemie, Technische Universität Dresden
Bergstrasse, 66b, 01062 Dresden (Germany)

[**] We are grateful to the help of R. Popovitz-Biro (HRTEM/EELS), Hilla Friedman (SEM/EDS), Y. Feldman (XRD), H. Cohen and T. Bendikov (XPS), S. R. Cohen (AFM), and Y. Tsverin (some of the conductivity measurements). R.T. gratefully acknowledges the support of ERC project INTIF 226639, the Israel Science Foundation, AddNano project 229284 of the FP7 (EU) program (late tribological measurements), the Harold Perlman Foundation, and the Irving and Cherna Moskowitz Center for Nano and Bio-Nano Imaging. R.T. is the Drake Family Chair in Nanotechnology and director of the Helen and Martin Kimmel Center for Nanoscale Science; A.I.F. and R.V. acknowledge support by the U.S. Department of Energy (DOE) Grant No. DE-FG02-03ER15476. Use of the NSLS was supported by the U.S. DOE Grant No. DE-AC02-98CH10886. Beamlines X18B and X19A at the NSLS are supported in part by the Synchrotron Catalysis Consortium, U. S. DOE Grant No DE-FG02-05ER15688.

 Supporting information for this article is available on the WWW under http://dx.doi.org/10.1002/anie.201105324.

between approximately 0.07–0.5 at % depending on the loading of the dopant, the temperature, and duration of the process. For the Re:IF-MoS$_2$ the Re level was found to increase with the formal ReO$_3$ content of the oxide powder and varied to some extent from one measurement to the other (three measurements for each Re concentration). These results suggested some non-uniformity in the doping level of the NP. Roughly speaking, the average Re concentration in the IF-MoS$_2$ powder was found to be about 1/3 to 1/2 of the formal (weighted) concentration in the oxide precursor. Thus in a typical nanoparticle consisting of about 5–7 × 10^5 Mo atoms, there are on the average about 200–300 Re atoms (in the 0.12 at % sample).

X-ray absorption fine structure (XAFS) measurements were performed to determine the dopant locations by checking the local structure around the Re atoms. Data processing and analysis of the raw X-ray absorption near edge structure (XANES) and extended XAFS (EXAFS) data were performed for IF-MoS$_2$, Re:IF-MoS$_2$ (0.12 and 0.71 at %). The Fourier transform magnitude of the data collected at the Re L$_3$ edge (Figure 2a) demonstrates distinct differences between the atomic arrangements of the Re-doped IF samples and bulk ReS$_2$. A theoretical model was constructed by replacing the X-ray-absorbing Mo atom in the MoS$_2$ structure by Re and using FEFF6 program to calculate the theoretical EXAFS contribution to Re from the first (Re–S) and second (Re–Mo) coordination shells. The good fit between the theoretical and experimental curves for the 0.12 at % sample (Figure 2b) support the substitution model for the dopant (Re$_{Mo}$).

Pure ReS$_2$ has a small Re–S first peak owing to the large first-shell disorder. Thus the larger peak intensity of the 0.12 at % Re compared to the 0.71 at % (Figure 2a) is

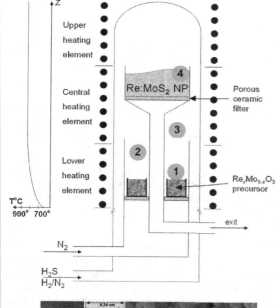

Figure 1. a) Modified quartz reactor used for the synthesis of the Re:IF-MoS$_2$ NP with the temperature profile shown along the reactor (z) axis, on the left, see text for details. b) HRSEM and HRTEM (inset) images of a typical Re(0.12 at%):IF-MoS$_2$ NP. The interlayer spacing as shown by the line profile of Re:IF-MoS$_2$ (0.627 nm) coincides with that of the undoped IF-MoS$_2$. c) HRTEM image of a post-synthesis Re-doped INT-WS$_2$.

distinction between this image and a typical TEM micrograph of pristine WS$_2$ nanotubes.

The Re concentration in the IF/INT was determined by inductive coupled plasma mass spectrometry (ICP-MS). For the Re-doped IF/INT-WS$_2$ the level of Re doping varied

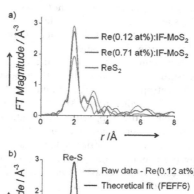
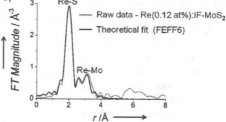

Figure 2. Fourier transform magnitudes of the EXAFS spectra of the Re L$_3$ edge for: a) Two Re-doped IF samples (0.12 at% red, 0.71 at% Re blue) and the ReS$_2$ (green) reference. b) Raw data (red) and theoretical fit (black) of the Re L$_3$ edge data in the 0.12 at% sample.

of the Harold Perlman Foundation, and the Irving and Cherna Moskowitz Center for Nano and Bio-Nano Imaging.

REFERENCES

(1) Tenne, R.; Margulis, L.; Genut, M.; Hodes, G. *Nature* **1992**, *360*, 444−446.
(2) Margulis, L.; Salitra, G.; Tenne, R.; Talianker, M. *Nature* **1993**, *365*, 113−114.
(3) Nath, M.; Rao, C. *J. Am. Chem. Soc.* **2001**, *123*, 4841−4842.
(4) Schuffenhauer, C.; Popovitz-Biro, R.; Tenne, R. *J. Mater. Chem.* **2002**, *12*, 1587−1591.
(5) Schuffenhauer, C.; Parkinson, B.; Jin Phillipp, N.; Joly Pottuz, L.; Martin, J.; Popovitz Biro, R.; Tenne, R. *Small* **2005**, *1*, 1100−1109.
(6) Popovitz-Biro, R.; Sallacan, N.; Tenne, R. *J. Mater. Chem.* **2003**, *13*, 1631−1634.
(7) Popovitz-Biro, R.; Twersky, A.; Hacohen, Y.; Tenne, R. *AIP Conf. Proc.* **2000**, *544*, 441−447.
(8) Levi, R.; Bar-Sadan, M.; Albu-Yaron, A.; Popovitz-Biro, R.; Houben, L.; Shahar, C.; Enyashin, A.; Seifert, G.; Prior, Y.; Tenne, R. *J. Am. Chem. Soc.* **2010**, *132*, 11214−11222.
(9) Albu Yaron, A.; Arad, T.; Popovitz Biro, R.; Bar Sadan, M.; Prior, Y.; Jansen, M.; Tenne, R. *Angew. Chem., Int. Ed.* **2005**, *44*, 4169−4172.
(10) Remskar, M.; Skraba, Z.; Cleton, F.; Sanjines, R.; Levy, F. *Appl. Phys. Lett.* **1996**, *69*, 351−353.
(11) Remskar, M.; Mrzel, A.; Virsek, M.; Godec, M.; Krause, M.; Kolitsch, A.; Singh, A.; Seabaugh, A. *Nanoscale Res. Lett.* **2010**, *6*, 26.
(12) Yella, A.; Mugnaioli, E.; Panthöfer, M.; Therese, H. A.; Kolb, U.; Tremel, W. *Angew. Chem., Int. Ed.* **2009**, *48*, 6426−6430.
(13) Radovsky, G.; Popovitz-Biro, R.; Staiger, M.; Gartsman, K.; Thomsen, C.; Lorenz, T.; Seifert, G.; Tenne, R. *Angew. Chem., Int. Ed.* **2011**, *50*, 12316−12320.
(14) Feldman, Y.; Frey, G.; Homyonfer, M.; Lyakhovitskaya, V.; Margulis, L.; Cohen, H.; Hodes, G.; Hutchison, J.; Tenne, R. *J. Am. Chem. Soc.* **1996**, *118*, 5362−5367.
(15) Feldman, Y.; Zak, A.; Popovitz-Biro, R.; Tenne, R. *Solid State Sci.* **2000**, *2*, 663−672.
(16) Zak, A.; Sallacan-Ecker, L.; Margolin, A.; Genut, M.; Tenne, R. *Nano* **2009**, *4*, 91−98.
(17) Magneli, A. *Ark. Kemi* **1950**, *1*, 513−523.
(18) Sahle, W. *J. Solid State Chem.* **1982**, *45*, 324−333.
(19) Zak, A.; Feldman, Y.; Alperovich, V.; Rosentsveig, R.; Tenne, R. *J. Am. Chem. Soc.* **2000**, *122*, 11108−11116.
(20) Feldman, Y.; Wasserman, E.; Srolovitz, D.; Tenne, R. *Science* **1995**, *267*, 222−225.
(21) Chen, J.; Li, S.-L.; Xu, Q.; Tanaka, K. *Chem. Commun.* **2002**, 1722−1723.
(22) Merchan-Merchan, W.; Saveliev, A. V.; Kennedy, L. A. *Chem. Phys. Lett.* **2006**, *422*, 72−77.
(23) Handa, H.; Abe, T.; Suemitsu, M. *e-J. Surf. Sci. Nanotechnol.* **2009**, *7*, 307−310.
(24) Cariati, F.; Bart, J. C. J.; Sgamellotti, A. *Inorg. Chim. Acta* **1981**, *48*, 97−103.
(25) Portemer, F.; Sundberg, M.; Kihlborg, L.; Figlarz, M. *J. Solid State Chem.* **1993**, *103*, 403−414.
(26) Frey, G. L.; Rothschild, A.; Sloan, J.; Rosentsveig, R.; Popovitz-Biro, R.; Tenne, R. *J. Solid State Chem.* **2001**, *162*, 300−314.
(27) Kihlborg, L. *Acta Chem. Scand.* **1960**, *14*, 1612−22.
(28) Yamazoe, N.; Ekstrom, T.; Kihlborg, L. *Acta Chem. Scand., Ser. A* **1975**, *29*, 404−8.
(29) Kihlborg, L. *Acta Chem. Scand.* **1969**, *23*, 1834−1835.
(30) Hu, J.; Odom, T.; Lieber, C. *Acc. Chem. Res.* **1999**, *32*, 435−445.
(31) Lauhon, L. J.; Gudiksen, M. S.; Lieber, C. M. *Philos. Trans. R. Soc. London, Ser. A* **2004**, *362*, 1247−1260.
(32) Bethune, D. S.; Klang, C. H.; de Vries, M. S.; Gorman, G.; Savoy, R.; Vazquez, J.; Beyers, R. *Nature* **1993**, *363*, 605−607.
(33) Hacohen, Y. R.; Popovitz-Biro, R.; Grunbaum, E.; Prior, Y.; Tenne, R. *Adv. Mater. (Weinheim, Ger.)* **2002**, *14*, 1075−1078.
(34) Iijima, S.; Ichihashi, T. *Nature* **1993**, *363*, 603−605.

(35) Thess, A.; Lee, R.; Nikolaev, P.; Dai, H.; Petit, P.; Robert, J.; Xu, C.; Lee, Y. H.; Kim, S. G.; Rinzler, A. G.; Colbert, D. T.; Scuseria, G. E.; Tománek, D.; Fischer, J. E.; Smalley, R. E. *Science* **1996**, *273*, 483−487.
(36) Wiesel, I.; Arbel, H.; Albu-Yaron, A.; Popovitz-Biro, R.; Gordon, J. M.; Feuermann, D.; Tenne, R. *Nano Res.* **2009**, *2*, 416−424.
(37) Afanasiev, P.; Geantet, C.; Thomazeau, C.; Jouget, B. *Chem. Commun.* **2000**, 1001−1002.
(38) Levy, M.; Albu Yaron, A.; Tenne, R.; Feuermann, D.; Katz, E. A.; Babai, D.; Gordon, J. M. *Isr. J. Chem.* **2010**, *50*, 417−425.
(39) Chibante, L. P. F.; Thess, A.; Alford, J. M.; Diener, M. D.; Smalley, R. E. *J. Phys. Chem.* **1993**, *97*, 8696−8700.
(40) Fields, C. L.; Pitts, J. R.; Hale, M. J.; Bingham, C.; Lewandowski, A.; King, D. E. *J. Phys. Chem.* **1993**, *97*, 8701−8702.
(41) Flamant, G.; Luxembourg, D.; Robert, J. F.; Laplaze, D. *Sol. Energy* **2004**, *77*, 73−80.
(42) Laplaze, D.; Bernier, P.; Flamant, G.; Lebrun, M.; Brunelle, A.; Della-Negra, S. *J. Phys. B: At., Mol. Opt. Phys.* **1996**, *29*, 4943−4954.
(43) Alvarez, L.; Guillard, T.; Sauvajol, J. L.; Flamant, G.; Laplaze, D. *Appl. Phys. A: Mater. Sci. Process.* **2000**, *70*, 169−173.
(44) Laplaze, D.; Bernier, P.; Maser, W. K.; Flamant, G.; Guillard, T.; Loiseau, A. *Carbon* **1998**, *36*, 685−688.
(45) Flamant, G.; Ferriere, A.; Laplaze, D.; Monty, C. *Sol. Energy* **1999**, *66*, 117−132.
(46) Albu Yaron, A.; Arad, T.; Levy, M.; Popovitz Biro, R.; Tenne, R.; Gordon, J. M.; Feuermann, D.; Katz, E. A.; Jansen, M.; Muehle, C. *Adv. Mater.* **2006**, *18*, 2993−2996.
(47) Albu-Yaron, A.; Levy, M.; Tenne, R.; Popovitz-Biro, R.; Weidenbach, M.; Bar-Sadan, M.; Houben, L.; Enyashin, A. N.; Seifert, G.; Feuermann, D.; Katz, E. A.; Gordon, J. M. *Angew. Chem., Int. Ed.* **2011**, *50*, 1810−1814.
(48) Flamant, G.; Robert, J. F.; Marty, S.; Gineste, J. M.; Giral, J.; Rivoire, B.; Laplaze, D. *Energy* **2004**, *29*, 801−809.

Chapter 3
PREAMBLE TO THE CHARACTERIZATION

In our early studies, and even now when we find a new kind of nanotube or fullerene-like nanoparticles in small amounts, transmission electron microscopy (TEM) is the main tool for characterization. Over the last two decades major developments in the field of electron microscopy, and TEM in particular, made this technique a full-fledged laboratory with immense possibilities for studying and characterizing nanostructures with unprecedented detail and accuracy. Obviously, in order to carry out detailed characterization studies, larger amounts i.e. mg quantities of the pure phases were needed. In the case of fullerene-like (IF) MoS_2 and WS_2 nanoparticles, such quantities became available in 1995 and 1996, respectively. I was also lucky to have Dr. Ronit Popovitz-Biro take charge of the TEM lab after the untimely death of Dr. Lev Margulis in 1995.

Thus, using X-ray diffraction, it became possible to show that the spacing between the layers along the c-axis increases by about 2% compared to the bulk phase (1.4, 2.1). This observation was attributed to the stress relaxation of the folded layers in these closed cage nanoparticles. The question of chirality of the different layers in a multiwall nanotube preoccupied our minds and is still an open issue. This issue is clearly related to the built-in strain in the folded layers. Theoretical analysis by Seifert, Tenne and their co-workers (2.6) indicated that multiwall MoS_2 nanotubes 5–10 layers thick are predominantly stable in the nano-regime and are semiconductors irrespective of their diameter and chirality. This analysis was experimentally confirmed in our work on WS_2 nanotubes [1].

Detailed optical spectroscopy measurements were carried out by my student Gitti Frey in the mid and late 90's. One of her important observations was that the IF nanoparticles of MoS_2 and WS_2 are semiconductors with excitonic structure, not much different from that of the bulk phase (3.1). Remarkably though, she observed a red shift in the A and the B excitons. When she first observed this red shift in 1994 we did not pay too much attention, and we attributed it to the structural defects in the IF nanoparticles. Nonetheless, when she repeated this kind of measurement for numerous samples and at different temperatures it became evident that a new phenomenon has been revealed. When studying the optical spectra of semiconductor nanoparticles one is bound to evoke the quantum size effect whereby the confinement

Scanning Tunneling Microscope Induced Crystallization of Fullerene-like MoS₂

M. Homyonfer,[†] Y. Mastai,[†] M. Hershfinkel,[‡] V. Volterra,[‡] J. L. Hutchison,[§] and R. Tenne*,[†]

Contribution from the Department of Materials and Interfaces, Weizmann Institute, Rehovot 76100, Israel, Department of Physics, Ben-Gurion University, P.O. Box 653, Beer-Sheva 84105, Israel, and Department of Materials, University of Oxford, Parks Road, Oxford OX1 3PH, U.K.

Received May 8, 1996[⊗]

Abstract: Amorphous precursors, like MoS_3 (WS_3), were shown before to be an ideal precursor for the growth of inorganic fullerene-like material in a rather slow crystallization process which lasts anything from 1 h at 800–900 °C[1,2] to a few years at ambient conditions.[3,4] Using a few microsecond short electrical pulses from the tip of a scanning tunneling microscope, crystallization of amorphous MoS_3 (a-MoS_3) nanoparticles, which were electrodeposited on a Au substrate into MoS_2 nanocrystallites with a fullerene-like structure (IF-MoS_2), is demonstrated. The (outer) shell of each nanocrystallite is complete, which suggests that the reaction extinguishes itself upon completion of the crystallization of the MoS_2 layers. A completely different mode of crystallization is observed in the case of continuous a-MoS_3 films. Here tiny (2–3 nm thick) 2H-MoS_2 platelets are observed after the electrical pulse, suggesting a very rapid dissipation of the thermal energy through the gold substrate, in the continuous domain. Since the reaction mechanism in both cases is believed to be the same, it is likely that the main stimulus for the chemical reaction/crystallization of the IF material results from the slow heat dissipation from the nanoparticle. The exothermicity of the chemical reaction may further promote the rate of the process.

Introduction

Nanoparticles of layered compounds are inherently unstable in the planar form and adopt polyhedral (fullerene-like) topologies upon crystallization. Conversion of metal oxide nanoparticles into fullerene-like MX_2 (M = W, Mo; X = S, Se) polyhedra and nanotubes has been demonstrated.[1–6] Spontaneous crystallization of hollow cage WS_2 polyhedra from amorphous WS_3 soot at room temperature was also observed within a few years time.[3,4] Irradiation of the sample with intense electron[3,7] and ion beams[8] accelerates the crystallization of fullerene and fullerene-like nanoparticles to a time scale measured in minutes. However, this is a rather aggressive process which must be carried out in a controlled atmosphere. The present report deals with the crystallization of fullerene-like MoS_2 nanocrystallites (IF-MoS_2) from an amorphous MoS_3 (a-MoS_3) precursor of nano-size dimensions at ambient conditions. A rather short electric pulse from a scanning tunneling microscope (STM) tip led to an abstraction of sulfur atoms and subsequent crystallization of fullerene-like nanoparticles. This process is attributed to the local heating of the nanoparticle, which is amplified by the small size of the contact area between the precursor nanoparticles and the gold substrate, and also to the exothermicity of the chemical reaction.

Experimental Section

For sample preparation, a 35 nm thick polycrystalline gold film was evaporated onto a quartz substrate at room temperature and then annealed for 12 h at 250 °C.[9] The textured film consisted of (111) oriented gold grains, ca. 0.25 μm in size. Subsequently, a discontinuous film consisting of a-MoS_3 nanoparticles was deposited on the gold substrate using a modification of a previous procedure.[10] The film was not uniform: in certain areas a-MoS_3 nanoparticles mostly having an oval shape were obtained, while in other areas a continuous film, consisting of Mo and S but with varying stoichiometry, was deposited. The film was prepared by electrochemical deposition from a bath of 0.1 M ammonium thiomolybdate and 0.05 M Na_2SO_4 dissolved in an aqueous solution at pH 5 and a bias of 0.23 V vs saturated calomel electrode (SCE).[10] The size of the nanocrystallites depended strongly on the bath temperature, with the typical size varying between 130, 40, and 5 nm at 10, 0, and −10 °C, respectively. The films which were used in the present work were prepared at 0 °C. Ambient scanning tunneling microscope (STM) provided with a Pt/Ir tip was used in the experiments. The surface of the sample was carefully scratched at certain points which served as markers for the identification of the STM-treated zones during the subsequent TEM analysis.

STM imaging of the a-MoS_3/Au surface was done with a tunneling current of about 0.5 nA and a positive tip bias of 1.5–2.7 V, compared with 0.2 V used for a pure gold film. Before applying the electric pulses, an area of ca. 50 × 50 nm² was defined on the sample. Pulses ranging between 4 and 9 V (tip positive) were applied to induce crystallization of the amorphous nanoparticles, while the tip was positioned a few angstroms above the sample. The duration of the pulses varied from 10 to 1000 μs. The system automatically stopped

---

[†] Weizmann Institute.
[‡] Ben-Gurion University.
[§] University of Oxford.
[⊗] Abstract published in *Advance ACS Abstracts*, August 1, 1996.
(1) Tenne, R.; Margulis, L.; Genut, M.; Hodes, G. *Nature* **1992**, *360*, 444.
(2) Margulis, L.; Salitra, G.; Tenne, R.; Talianker, M. *Nature* **1993**, *365*, 113.
(3) Hershfinkel, M.; Gheber, L. A.; Volterra, V.; Hutchison, J. L.; Margulis, L.; Tenne, R. *J. Am. Chem. Soc.* **1994**, *116*, 1914.
(4) Margulis, L.; Tenne, R.; Iijima, S. *Microscopy, Microanalysis and Microstrucutres*; in press.
(5) Feldman, Y.; Wasserman, E.; Srolovitz, D. J.; Tenne, R. *Science* **1995**, *267*, 222.
(6) Feldman, Y.; Frey, G. L.; Homyonfer, M.; Lyakhovitskaya, V.; Margulis, L.; Cohen, H.; Hodes, G.; Hutchison, J. L.; Tenne, R. *J. Am. Chem. Soc.* in press.
(7) Ugarte, D. *Nature* **1992**, *359*, 707.
(8) Chadderton, L. T.; Fink, D.; Gamaly, Y.; Moeckel, H.; Wang, L.; Omichi, H.; Hosoi, F. *Nucl. Instrum. Methods Phys. Res.* **1994**, *B91*, 71.
(9) Golan, Y.; Margulis, L.; Rubinstein, I. *Surf. Sci.* **1992**, *264*, 312.
(10) Bélanger, D.; Laperrière, G.; Marsan, B. *J. Electoanal. Chem.* **1993**, *347*, 165.

© 1996 American Chemical Society

the scanning prior to the pulse. During the pulse, the feedback loop was disengaged and a sample-hold loop fixed the tip in its position. After the pulse was applied, the feedback system was reactivated and the scanning was resumed. The predefined area was scanned repetitively to verify that apart from the oblate objects which appeared as a result of the pulse, no other modifications on the sample surface had occurred. Usually a few series of scan/pulse/scan routines were applied for the same predefined area. Subsequently, another 50 × 50 nm^2 area was defined on the same sample and the same procedure was repeated, and so forth. In several cases, well-defined nanoscopic features appeared following the pulses while in other cases no change in the STM image was noticed after a series of pulses. This irregularity is not surprising in view of the heterogeneity of the sample surface as pointed out above. Quite often, a pulse caused an instability in the successive scans. This instability was attributed to a cluster of sulfur atoms, which detached from the a-MoS$_3$ particle (film) during crystallization and stuck to the tip (*vide infra*).

A Philips model EM400 transmission electron microscope (TEM) with beam voltage of 100 keV was used for the analysis of most of the samples, while a high-resolution transmission electron microscope (HRTEM) JEOL model 2010 (beam voltage 200 keV) equipped with an energy dispersive analyzer (EDS) Link model ISIS was used for imaging and compositional analysis of a few samples (±1% accuracy) only. The size of the EDS probe was varied between 3 and 5 nm and the takeoff angle of the detector was changed. The results of the quantitative analysis were not influenced by varying the size of the probed zone or the takeoff angle. In order to peel-off the sample from the substrate without damaging it or the markers and transfer it onto the TEM grids, the sample edges where dipped briefly into a dilute (ca. 10%) HF solution. As soon as the edges of the thin gold film detached from the quartz, it was lifted-off completely very carefully and put on a grid with alpha-numerical markers, so that the STM treated zone could be easily identified under the TEM (HRTEM). This procedure permitted exact identification of the STM-treated zone without any difficulty. Prior to that, an alternative procedure was used according to which the identification of the treated zone was tedious and required prolonged examination of the sample with TEM/HRTEM, and was furthermore not always successful. The atomic percent content of each nanoparticle, which was based on the data collected by the EDS analyzer, was determined using the quantitative Cliff–Lorimer method for thin sections.

Results

Since a-MoS$_3$ is insulating, STM imaging of the nanoparticles or the continuous film was not possible. For this reason, TEM and later on HRTEM were the most useful means for direct imaging of the nanoparticles and the film. Following the electrical pulse, typically 20 nm round objects were identified by the STM (Figure 1). The contour of the STM tip across such a nanoparticle was quite spherical indeed. Atomically resolved images of these objects could not be obtained with the STM, due likely to the curvature of the objects which implies that tunneling from more than one atom of the STM tip to the sample (and *vise versa*) takes place. The STM images resemble the ones reported before,[3] but it was not possible to decide whether the synthesis yielded nanoparticles with fullerene-like structure, at this stage. However, the appearance of such objects after the electrical pulses served as a criterion for the success of such experiments and only such samples were further analyzed by TEM, etc. To get a better idea of the structure of the nanoparticles, the treated sample was carefully transferred onto a polymer coated Cu grid and a TEM analysis was carried out. Figure 2a shows a TEM image of a group of amorphous oval a-MoS$_3$ nanoparticles deposited on the textured gold film. The average size of the nanoparticles is approximately 30 nm. Smaller a-MoS$_3$ nanoparticles are often observed (see inset of Figure 2a). A TEM image of a typical group of fullerene-like IF-MoS$_2$ particles, which were obtained after a series of electrical pulses were applied to the original film, is shown in

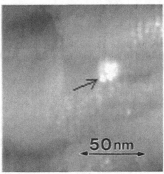

Figure 1. STM images of assortments of IF-MoS$_2$ nanoparticles produced by electrical pulse from the tip of the STM. Pulse voltage was 5 V in these experiments. The nature of each of the IF particles was verified through the semiconductor behavior of the I–V curve.

Figure 2b. Note that in most cases, the core of the nanoparticles remained amorphous. In the inset of Figure 2b a 15-nm nanoparticle is shown, which has been fully converted from a-MoS$_3$ into IF-MoS$_2$. Growth fronts of the crystalline phase inside the amorphous core can be discerned (marked by arrows). Pulse experiments on top of the continuous nonstoichiometric Mo–S film or with pulse voltage smaller than 4 V did not lead to crystallization of fullerene-like structures (see also refs 11 and 12). Contrarily, pulses of 5 V and above produced fullerene-like objects, reproducibly. Complete destruction of the film surface and usually loss of an image was obtained with a pulse voltage larger than 8 V. The topology of the nested fullerene-like objects resembled structures reported earlier.[1-6]

Contrary to this behavior, a completely different mode of crystallization was observed for areas which were covered by

---

(11) Schimmel, Th.; Fuchs, H.; Lux-Steiner, M. *Phys. Status Solidi A* **1992**, *131*, 47.

(12) Akari, S.; Möller, S.; Dransfeld, K. *Appl. Phys. Lett.* **1991**, *59*, 243.

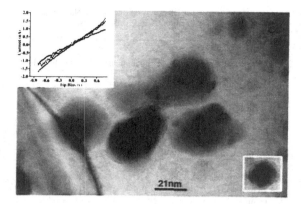

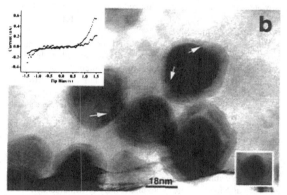

Figure 2. (a) TEM image of a group of a-MoS$_3$ nanoparticles. In the insets are shown an amorphous particle of ca. 15 nm and a typical I−V curve obtained with this film. (b) Crystalline IF-MoS$_2$ nanoparticles with an a-MoS$_3$ core. The insets show a fullerene-like nanoparticle of ca. 15 nm and a typical I−V curve obtained while the tip was fixed a few angstroms above one of the nanoparticles.

a continuous thin film of a-MoS$_3$. As shown in Figure 3, small 2H-MoS$_2$ platelets, which were embedded in the amorphous matrix, were formed as a result of the electrical pulse. The size of the crystalline domain is rather small (2−3 nm thick) in this case. The dissipation of the thermal energy, deposited by the electrical pulse in the continuous Mo−S film, was probably very fast and hence the energy density was below the required threshold for sulfur abstraction and crystallization of the material. Attempts to use the same method to obtain IF structures from large area MoS$_2$ crystals were not successful. This result clearly indicates that IF-MoS$_2$ could occur only in discontinuous nanoscopic a-MoS$_3$ domains of a diameter smaller than ca. 50 nm.

Prior to the application of the electrical pulses, an Ohmic behavior was observed for the film (inset of Figure 2a). The I−V curve, which was measured while the tip was fixed above fullerene-like nanoparticles (inset of Figure 2b), exhibited a typical semiconductor-like behavior with a statistically averaged bandgap of −1.1 eV (±0.05 eV).[3] The value of the bandgap is ca. 0.1 eV smaller than the bandgap of the bulk (2H) material and was confirmed by direct optical measurements.[13] The shape of the I−V curve, which changed from a metallic-like behavior to a semiconductor-like behavior, was another criteria used to

(13) Frey, G. L.; Homyonfer, M.; Feldman, Y.; Tenne, R. To be submitted for publication.

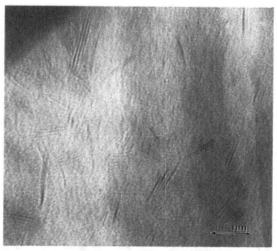

Figure 3. TEM image of a continuous nonstoichiometric Mo−S film on which a series of a few electrical pulses were applied and led to the crystallization of 2H-MoS$_2$ nano-size (ca. 2−3 nm) platelets.

judge the success of an experiment. The van der Waals surface of MoS$_2$ is known to be very inert against oxidation, which is probably the reason for the fact that atomic resolution for the 2H polytype and meaningful values for the bandgap can be reproducibly obtained with ambient STM.[3]

Unfortunately, it is not possible to determine the accurate tip position with respect to the nanoparticle, while the electrical pulse was applied. Furthermore, the size of the IF particles did not seem to depend on the pulse potential (between 5 and 8 V), which indicates that their size is determined by the size of the precursor (a-MoS$_3$) nanoparticles. This observation is not unique to the present process, since the size of the IF particles, which were produced from an oxide precursor, was also found to depend solely on the size of the precursor nanoparticles.[6] This analogy is not surprising in view of the fact that in both cases the crystallization goes from the outside to the inside.

In order to get more information about the process, HRTEM/EDS were used to analyze the samples prior to and after the electrical pulses. The high resolution of the microscope, coupled with the small thickness of the film and substrate, permitted chemical analysis of the nanoparticles one by one. Figure 4a shows a typical EDS spectrum taken from a single a-MoS$_3$ nanoparticle, while Figure 4b shows a typical EDS spectrum of a single IF-MoS$_2$ particle. Apart from the contributions of the gold substrate, the Cu grid, and the constituting elements of the nanoparticles, no other foreign element could be detected. Table 1 shows the calculated atomic ratios of the various elements of the two typical nanoparticles. While the ratio of S/Mo was found to be 3.0 for the amorphous nanoparticles, the ratio varied between 1.8 and 2.2 for the nanoparticles with IF structure, in accordance with the hypothesized mechanism. Note that the contribution of the Cu grid stems from the large cross section of this metal with respect to X-ray emission.

Modifications of materials by pulse from the STM tip have been published.[15−18] The main mechanisms proposed for the process are field emission of electrons and atoms, deposition

(14) Srolovitz, D. J.; Safran, S. A.; Homyonfer, M.; Tenne, R. *Phys. Rev. Lett.* **1995**, *74*, 1799.
(15) Staufer, U.; Wiesendanger, R.; Eng, L.; Rosenthaler, L.; Hidber, H. R.; Güntherodt, H.-J.; Garcia, N. *Appl. Phys. Lett.* **1987**, *51*, 244.

signed to the Raman inactive A_{2u} mode, which is activated by the strong resonance effect. Dynamic band calculations show that the coupling of the A_{1g} mode to the direct electronic transition is responsible for the enhancement of the A_{1g} mode under resonance conditions. Finally, studying the position of the dispersive band at 429 cm^{-1} in several MoS$_2$ samples with different band gaps showed a softening of the Raman inactive B_{2g}^2 mode. Since the quasiacoustic B_{2g}^2 mode involves the vibration of S-Mo-S planes against each other, the slight softening of this mode is consistent with the 2–4% c-axis lattice expansion previously observed in IF-MoS$_2$ by XRD measurements.[10] Consequently, we expect the E_{2g}^2 rigid-layer mode in the IF material, which is currently under investigation, to show the same trend and to shift to lower energies.

ACKNOWLEDGMENTS

G.L.F. and R.T. wish to thank Dr. Philip Klipstein for helpful discussions. This work was supported by the Minerva Foundation (Munich), ACS-PRF (USA), US-Israel Binational Foundation, and Applied Materials-Weizmann Foundation. M.J.M., M.S.D., and G.D. acknowledge support from NEDO and from NSF Contract No. DMR 98-04734.

---

[1] A. P. Alivisatos, J. Phys. C **100**, 13 226 (1996).

[2] C. B. Murray, D. J. Norris, and M. G. Bawendi, J. Am. Chem. Soc. **115**, 8706 (1993).

[3] P. M. Ajayan and T. W. Ebbesen, Rep. Prog. Phys. **60**, 1025 (1997).

[4] R. Tenne, L. Margulis, M. Genut, and G. Hodes, Nature (London) **360**, 444 (1992).

[5] L. Margulis, G. Salltra, and R. Tenne, Nature (London) **365**, 114 (1993).

[6] M. Remskar, Z. Skraba, F. Cléton, R. Sanjinés, and F. Lévy, Appl. Phys. Lett. **69**, 351 (1996).

[7] M. Remskar, Z. Skraba, M. Regula, C. Ballif, R. Sanjinés, and F. Lévy, Adv. Mater. **10**, 246 (1998).

[8] P. A. Parilla, A. C. Dillon, K. M. Jones, G. Riker, D. L. Schulz, D. S. Ginley, and M. J. Heben, Nature (London) **397**, 114 (1999).

[9] L. Rapoport, Y. Bilik, Y. Feldman, M. Homyonfer, S. R. Cohen, and R. Tenne, Nature (London) **387**, 791 (1997).

[10] Y. Feldman, E. Wasserman, D. J. Srolovitz, and R. Tenne, Science **267**, 222 (1995).

[11] M. Hershfinkel L. A. Gheber, V. Voltera, J. L. Hutchison, L. Margulis, and R. Tenne, J. Am. Chem. Soc. **116**, 1914 (1994).

[12] Y. Feldman, G. L. Frey, M. Homyonfer, V. Lyakhovitskaya, L. Margulis, H. Cohen, G. Hodes, J. L. Hutchison, and R. Tenne, J. Am. Chem. Soc. **118**, 5362 (1996).

[13] M. Homyonfer, B. Alperson, Y. Rosenberg, L. Sapir, H. Cohen, G. Hodes, and R. Tenne, J. Am. Chem. Soc. **119**, 2693 (1997).

[14] G. L. Frey, S. Ilani, M. Homyonfer, Y. Feldman, and R. Tenne, Phys. Rev. B **57**, 6666 (1998).

[15] R. Coehoorn, C. Hass, and R. A. de Groot, Phys. Rev. B **35**, 6203 (1987).

[16] J. V. Acrivos, W. Y. Liang, J. A. Wilson, and A. D. Yoffe, J. Phys. C **4**, L18 (1971).

[17] J. P. Wilcoxon and G. A. Samara, Phys. Rev. B **51**, 7299 (1995).

[18] T. J. Wieting and J. L. Verble, Phys. Rev. B **3**, 4286 (1971).

[19] J. M. Chen and C. S. Wang, Solid State Commun. **14**, 857 (1974).

[20] T. Sekine, K. Uchinokura, T. Nakashizu, E. Matsuura, and R. Yoshizaki, J. Phys. Soc. Jpn. **53**, 811 (1984).

[21] A. M. Stacy and D. T. Hodul, J. Phys. Chem. Solids **46**, 405 (1985).

[22] C. Sourisseau, F. Cruege, and M. Fouassier, Chem. Phys. **150**, 281 (1991).

[23] R. Hoffman, J. Chem. Phys. **39**, 1397 (1963).

[24] M. Brändle, G. Calzaferri, and M. Lanz, Chem. Phys. **201**, 141 (1995). The program was downloaded from the World Wide Web homepage of Professor G. Calzaferri at the University of Bern (http://iacrs1.unibe.ch/members/biconcedit.html).

[25] D. J. Srolovitz, S. A. Safran, M. Homyonfer, and R. Tenne, Phys. Rev. Lett. **74**, 1779 (1995).

[26] M. J. Lipp, V. G. Baonza, W. J. Evans, and H. E. Lorenzana, Phys. Rev. B **56**, 5978 (1997).

[27] X. S. Zhao, Y. R. Ge, J. Schroeder, and P. D. Persans, Appl. Phys. Lett. **65**, 2033 (1994).

[28] H. Richter, Z. P. Wang, and L. Ley, Solid State Commun. **39**, 625 (1981).

[29] T. Kanata, H. Murai, and K. Kubota, J. Appl. Phys. **61**, 969 (1987).

[30] W. H. Weber, K. C. Hass, and J. R. McBride, Phys. Rev. B **48**, 178 (1993).

[31] N. Wakabayashi, H. G. Smith, and R. M. Nicklow, Phys. Rev. B **12**, 659 (1975).

[32] M. Cardona, in *Light Scattering in Solids II*, edited by M. Cardona and G. Guntherödt (Springer-Verlag, Berlin, 1983), p. 19.

[33] S. Jimenez Sandoval, D. Yang, R. F. Frindt, and J. C. Irwin, Phys. Rev. B **44**, 3955 (1991).

[34] F. Tuinstra and J. L. Koenig, J. Chem. Phys. **53**, 1126 (1970).

[35] A. V. Baranov, A. N. Bekhterev, Ya. S. Bobovich, and V. I. Petrov, Opt. Spectrosk. **62**, 1036 (1987) [Opt. Spectrosc. **62**, 612 (1987)].

[36] R. J. Nemanich and S. A. Solin, Phys. Rev. B **20**, 392 (1979).

[37] R. Srivastava and L. L. Chase, Solid State Commun. **11**, 349 (1972).

[38] J. L. Verble and T. J. Wieting, Phys. Rev. Lett. **25**, 362 (1970).

[39] T. J. Wieting and A. D. Yoffe, Phys. Status Solidi **37**, 353 (1970).

[40] B. L. Evans and P. A. Young, Phys. Status Solidi **25**, 417 (1968).

[41] A. G. Bagnall, W. Y. Liang, E. A. Marseglia, and B. Welber, Physica B **99**, 343 (1980).

Alkali Metal Intercalated Fullerene-Like MS$_2$ (M = W, Mo) Nanoparticles and Their Properties

Alla Zak,[†] Yishay Feldman,[‡] Vera Lyakhovitskaya,[†] Gregory Leitus,[†] Ronit Popovitz-Biro,[†] Ellen Wachtel,[‡] Hagai Cohen,[‡] Shimon Reich,[†] and Reshef Tenne*,[†]

Contribution from the Department of Materials and Interfaces, Weizmann Institute, Rehovot 76100, Israel, and Chemical Services Unit, Weizmann Institute, Rehovot 76100, Israel

Received August 28, 2001

Abstract: Layered metal disulfides-MS$_2$ (M = Mo, W) in the form of fullerene-like nanoparticles and in the form of platelets (crystallites of the 2H polytype) have been intercalated by exposure to alkali metal (potassium and sodium) vapor using a two-zone transport method. The composition of the intercalated systems was established using X-ray energy dispersive spectrometer and X-ray photoelectron spectroscopy (XPS). The alkali metal concentration in the host lattice was found to depend on the kind of sample and the experimental conditions. Furthermore, an inhomogeneity of the intercalated samples was observed. The product consisted of both nonintercalated and intercalated phases. X-ray diffraction analysis and transmission electron microscopy of the samples, which were not exposed to the ambient atmosphere, showed that they suffered little change in their lattice parameters. On the other hand, after exposure to ambient atmosphere, substantial increase in the interplanar spacing (3–5 Å) was observed for the intercalated phases. Insertion of one to two water molecules per intercalated metal atom was suggested as a possible explanation for this large expansion along the c-axis. Deintercalation of the hydrated alkali atoms and restacking of the MS$_2$ layers was observed in all the samples after prolonged exposure to the atmosphere. Electric field induced deintercalation of the alkali metal atoms from the host lattice was also observed by means of the XPS technique. Magnetic moment measurements for all the samples indicate a diamagnetic to paramagnetic transition after intercalation. Measurements of the transport properties reveal a semiconductor to metal transition for the heavily K intercalated 2H-MoS$_2$. Other samples show several orders of magnitude decrease in resistivity and two- to five-fold decrease in activation energies upon intercalation. These modifications are believed to occur via charge transfer from the alkali metal to the conduction band of the host lattice. Recovery of the pristine compound properties (diamagnetism and semiconductivity) was observed as a result of deintercalation.

Introduction

Transition metal dichalcogenides MS$_2$ (M = Mo, W) and their intercalated complexes belong to a large class of the so-called two-dimensional or layered solids. The layers of these materials consist of three interconnected, hexagonally arranged, atomic sheets (S-M-S). Trigonal prismatic or octahedral coordination of M to the S atoms occurs within each layer. In MoS$_2$ and WS$_2$ variation in the stacking sequence of the layers can lead to the formation of either a hexagonal polymorph with two layers in the unit cell (2H), rhombohedral with three layers (3R), or trigonal with one layer (1T). Atoms within a layer are bound by strong covalent forces, while individual layers are held together by van der Waals (vdW) interactions. The weak interlayer vdW interactions offer the possibility of introducing foreign atoms or molecules between the layers, e.g., via intercalation. By variation of the intercalant and its concentration, a large number of compounds with different properties can be prepared.

A number of experimental methods were used for the metal intercalation into the host MS$_2$ lattice: (1) immersion of the crystals or powders in metal-ammonia solution or in a solution of butyl-lithium in hexane; (2) exfoliation and subsequent restacking of the suspension of single layer MS$_2$ around the guest; (3) exposure to metal vapor, and (4) through an electrochemical process. Each method has its pros and cons. In particular, wet (from solution) processes provide very good control of the amount of intercalated atoms, but they usually lead to the co-intercalation of the solvent molecules.

The intercalation mechanism has been elucidated in a number of studies. It has been shown that intercalation of transition metal dichalcogenides by molecules such as ammonia,[1] or alkali atoms[2] develops first by adsorption of the molecules on the outer crystallite surface. Subsequently, a delay period follows, which appears to be associated with the weakening of the vdW forces between the top transition metal dichalcogenide layers. At the

---

* To whom correspondence should be addressed.
[†] Department of Materials and Interfaces, Weizmann Institute.
[‡] Chemical Services Unit, Weizmann Institute.

(1) (a) Acrivos, J. V. In *Intercalated Layered Materials*; Levi, F., Ed.; D. Reidel Publishing Company: Dordrecht, 1979; Vol. 6, Chapter 1, p 33; b) Beal, A. R.; Acrivos. J. V. *Philos. Mag. B.* **1978**, *37*, 409.
(2) Scrolz, G. A.; Joensen, P.; Reyes, J. M.; Frindt, R. F. *Physica* **1981**, *105B*, 214.

same time, a diffusion of the adsorbed, activated molecules around and in through the edges into these interlayer spaces is followed by weakening of the next layers down and so on. In this way, the elastic energy required to cause the lattice c-axis expansion is minimized. An alternative mechanism, where diffusion of the intercalated atoms occurs through the layers was established in the case of Cu on SnS_2[3] and proposed for K on WS_2,[4] both intercalated by exposure to the corresponding metal vapor.

Two main effects were identified as a result of the intercalation: the first — expansion of the interlayer spacing, which can approach 60 Å for organic molecule intercalation complexes;[5] the second — charge transfer from the intercalant to the host material.

Moreover, the tendency for charge transfer from a guest molecule to the host lattice appears to be the driving force for the intercalation reaction. Charge transfer can change the electronic properties of the material, raising the Fermi level, E_F, and increasing the free electron concentration by a few orders of magnitude. Usually, the lowest lying unoccupied energy levels in the host layers are derived from the transition metal d bands. The increase in d band filling of the host material provides the means for a controlled variation of many of its physical properties over a wide range. It is thus possible to achieve semiconductor-to-metal and metal-to-superconductor transitions by intercalation. Thus, semiconducting $2H-MoS_2$ intercalated with alkali metals was shown to exhibit metallic behavior and superconductivity with transition temperatures in the range of 3.7 to 6.3 K.[6] Potassium intercalated $2H-WS_2$ demonstrates a semiconductor-to-metal transition.[4] The intercalation process can be easily reversed (upon exposure of the sample to air, for example) and the "pure" host compound is recovered. This is an important characteristic for application of these materials as electrodes in secondary (rechargeable) batteries.

The data for lattice expansion and crystallographic transformation in the intercalated complexes is reported and reviewed here in some extent. The immersion of the crystals in alkali metal-ammonia solution for intercalation has been taken first by Rudorff[7] and Somoano et al.[6] resulting in partial intercalation of $2H-MoS_2$ with the following stoichiometries: K_xMoS_2 and Na_yMoS_2, $x = 0.4-0.6$ and $y = 0.3-0.6$. X-ray data of the intercalated MoS_2 samples indicated that, the unit cell was expanded upon intercalation, primarily in the c- direction. Two different expansions were found by Somoano[6] et al. in sodium and potassium intercalated samples, while only one expansion was found by Rudorff.[7] Thus, K-intercalated materials exhibit the lattice expansion of 4.25 / 5.73 Å[6] or 1.95 Å[7] and Na-intercalated materials 1.59 / 2.65 Å[6] or 1.35 Å.[7] The discrepancy between the lattice parameters of the same system, observed in the different works, can be probably attributed to the insertion of ammonia molecules (solvent) or water uptake from the environment by the intercalated metal atoms during or shortly after the process.

The water uptake (hydration) by intercalated alkali metals was discussed in detail by Wypych et al.[8-10] The alkali metal intercalated materials were found to undergo hydration and partial oxidation by washing with water to form $K_{1-x}(H_2O)_yMoS_2$. It was shown that the spacing along the c-axis of the intercalated crystals varies during the hydration reaction.

The K- and Na-intercalation of MS_2 (M = Mo, W) by the exfoliation/restacking technique resulted in interlamellar d spacing in the range of 9.3–9.7 Å ($\Delta c / 2 = 3.1$–3.5 Å expansion), which was attributed to co-intercalation of approximately one monolayer of water molecules.[11] $2H-MoS_2$ intercalated with Mn, Fe, Co, and Ni by the same technique reveals interlayer spacing expansion ($\Delta c / 2$) in the range of 5.1–5.3 Å, which is attributed to the co-intercalation of the metal atom together with two layers of water molecules.[12,13] Solid-state NMR study of the hydrated alkali-metal intercalated compounds of MoS_2 indicates on similar different expansions due to the different degree of hydration.[14]

Py et al.,[15] who have used the butyl-lithium technique for the metal intercalation, show that in addition to lattice expansion, the structure of $2H-MS_2$ may partially distort to $1T-MS_2$ upon intercalation. Later on, it was reported[16] that for low Li content ($x = 0.35$ in Li_xWS_2) the diffraction pattern was identical to that of $2H-WS_2$, with no change in basal plane spacing. However, for high lithium content ($x > 1$), the diffraction pattern shows a crystallographic transformation, where the transition metal coordination changes from trigonal prismatic (2H) to octahedral (1T). For moderate Li content ($x = 0.8$), a mixed phase of 2H and 1T was observed. At the same time, the c-lattice constant expands. Moreover, the 1T polytype has a metallic character. However, the distorted $1T-MS_2$ phase is known to be metastable and transforms to the thermodynamically stable $2H-MoS_2$ upon heating (above 90 °C) or aging in air, and simultaneously the guest species are reduced and deintercalated.[13] Different structures also were found, depending on the thermal history of each sample.

Some intercalation compounds (graphite, for example)[5,17-19] exhibit a high degree of ordering, which is revealed in a staging phenomenon. Ordering of the guest atoms (staging) in intercalation complexes of transition metal dichalcogenides has been observed only with certain concentrations of intercalant atoms and at low temperatures (Ag in TiS_2, for example).[17]

Nanoparticles of layered compounds (graphite, transition metal dichalcogenide, BN, $NiCl_2$, etc.) were shown to be unstable against folding and they close into fullerenes or fullerene-related structures (single or multiwall polyhedra, and

---

(3) (a) Ohuchi, F. S.; Jaegermann, W.; Parkinson, B. A. *Surf. Sci.* **1988**, *194*, L69; b) Jaegermann, W.; Ohuchi, F. S.; Parkinson, B. A. *SIA Surf. Interface Anal.* **1988**, *12*, 293.
(4) Ohuchi, F. S.; Jaegermann, W.; Pettenkofer, C.; Parkinson, B. A. *J. Am. Chem. Soc.* **1989**, *5*, 439.
(5) Liang, W. Y. In *Intercalation in Layered Materials*; Dresselhaus, M. S., Ed.; NATO ASI Series B: Physics, Plenum Press: New York, 1986; Vol. 148, p 31.
(6) Somoano, R. B.; Hadek, V.; Rembaum, A. *J. Chem. Phys.* **1973**, *58*, 697.
(7) Rudorff W. *Chimia* **1965**, *19*, 489.
(8) Wypych, F.; Schollhorn, R. *J. Chem. Soc., Chem. Commun.* **1992**, 1386.
(9) Wypych, F.; Weber, Th.; Prins, R. *Surf. Sci.* **1997**, *380*, L474-L478.
(10) Wypych, F.; Solenthaler, C.; Prins, R.; Weber, Th. *J. Solid State Chem.* **1999**, *144*, 430.
(11) Heising, J.; Kanatzidis, M. G. *J. Am. Chem. Soc.* **1999**, *121*, 11720.
(12) Zubavichus, Y. V.; Slovokhotov, Y. L.; Schilling, P. J.; Tittsworth, R. C.; Golub, A. S.; Protzenko, G. A.; Novikov, Y. N. *Inorganica Chimica Acta* **1998**, *280*, 211.
(13) Dungey, K. E.; Curtis, M. D.; Penner-Hahn, J. E. *Chem. Mater.* **1998**, *10*, 2152.
(14) Alexiev, V., Meyer, H. zu Altenschildesche, Prins, R., Weber, Th. *Chem. Mater.* **1999**, *11*, 1. 1742–1746..
(15) Py, M. A.; Haering, R. R. *Can. J. Phys.* **1983**, *61*, 76.
(16) Yang, D.; Frindt, R. F. *J. Phys. Chem. Solids* **1996**, *57*(6–8), 1113–1116.
(17) Brec, R. and Rouxel, J.; Brec, R. In *Intercalation in Layered Materials*; Dresselhaus, M. S., Ed.; NATO ASI Series B: Physics, Plenum Press: New York, 1986; Vol. 148, 31–125.

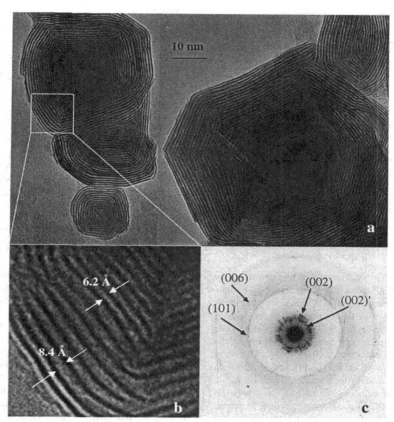

Figure 7. (a) TEM micrograph of the K-intercalated *IF*-MoS$_2$ nanoparticles after one week exposure to ambient atmosphere; (b) exploded view of the frame and (c) electron diffraction of (a).

Table 3. Summary of the XPS Data

	Alkali metal:Mo(W) atomic ratio			
	+K		+Na	
samples	as received	after neutralizer[a] (surface enrichment-%)[b]	as received	after neutralizer[a] (surface enrichment-%)[b]
2H-MoS$_2$	1:2.5	1:1 (150)	1:0.93[c]	1:0.51 (82)[c]
IF-MoS$_2$	1:8	1:6 (33)	1:9	1:8 (10)
IF-WS$_2$	1:17	1:9 (90)	1:13	1:13 (1)

[a] After a 15-h irradiation of the flood-gun on the sample. [b] Increase in the concentration of the alkali metal at the surface induced by irradiation with the flood-gun (deintercalation). [c] These data indicate the presence of Na on the surface of the intercalated material (Na:Mo>1) as a result of alkali metal excess during the intercalation. However, the increase in the concentration of the alkali metal at the surface induced by strong electric field with the flood-gun (deintercalation) provides strong evidence for the existence of guest atoms inside the crystal matrix.

formed across the particles. A significant surface enrichment of the alkali metal concentration occurs along a typical 5−20 h period, as indicated in Table 3. It is believed that the ionic nature of the alkali metals plays an important role in this type of 'dry electrochemistry'. The positive alkali metal ions diffuse out from the host lattice toward the extra negative charge induced on the surface by the electron beam flux. This out-diffusion process could not be obtained with low bias voltages of the flood-gun. This fact suggests that a critical field is needed for the deintercalation. Furthermore, the efficiency of the deintercalation process varies between the different samples. It is clearly much higher in the 2H matrixes as compared to their *IF* analogues. Also, K is found to outdiffuse from the host lattice much faster than Na.

5. Transport and Magnetic Properties. Transport and magnetic properties measurements gave additional evidence for the successful intercalation process. These measurements were carried out on the partially intercalated materials consisting of a mixture of nonintercated and intercalated phases. The 2H-MS$_2$ (M = Mo, W) materials are semiconductors[40,41] with a gap between the filled d$_{z^2}$ subband (the top of the valence band) and the conduction band based on the higher-lying d$_{x^2-y^2}$ and d$_{xy}$ orbitals. Upon intercalation an ionization (K → K$^+$ + e$^-$) of the intercalant occurs, resulting in donation of electrons into the lowest empty energy band.

5.1. Magnetic Susceptibility Measurements (MSM). MSM were performed for all four kinds of the materials (2H-MS$_2$ and *IF*-MS$_2$ (M = Mo, W)), prior to and after intercalation with both kinds of intercalating atoms (K, Na). However, only the data for K-intercalated *IF*-WS$_2$ is presented here.

Tungsten and molybdenum disulfides with trigonal prismatic coordination of metal atoms are diamagnetic,[40] which was confirmed by the present measurements (not shown). Magnetic susceptibility of -10^{-7} to -10^{-6} emu/g was obtained for these materials. The original bulk diamagnetism is preserved in the

(40) Wilson, J. A.; Yoffe A. D. *Adv. Phys.* **1969**, *18*, 193.
(41) Friend, R. H.; Yoffe, A. D. *Adv. Phys.* **1987**, *36*, 1.

Properties of Fullerene-Like MS$_2$ Nanoparticles

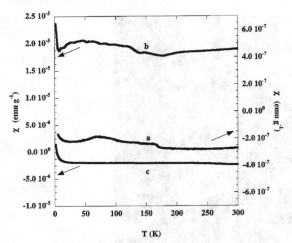

Figure 8. Magnetic susceptibility obtained at 100 Oe (a) for pristine *IF*-WS$_2$; (b) for K-intercalated *IF*-WS$_2$; (c) for deintercalated *IF*-WS$_2$.

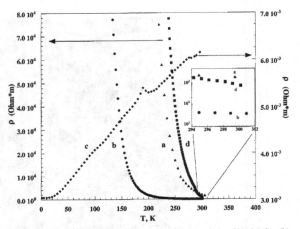

Figure 9. Resistivity versus temperature for (a) pristine 2H−MoS$_2$, (b) K-intercalated 2H−MoS$_2$ with 12 at. % of intercalant; (c) K− intercalated 2H−MoS$_2$ with 18 at. % of intercalant and (d) for deintercalated c-sample.

fullerene-like nanoparticles, but nevertheless two new magnetic transitions are revealed at 60 and 160 K, as shown in Figures 8a. Similar transition occurs at 60 K also for the oxide nanoparticles, which serve as the precursors for the *IF*-MS$_2$ synthesis. However, single-crystal WO$_3$ is diamagnetic, while MoO$_3$ is a very weak paramagnet[42] and they do not exhibit the aforementioned magnetic transitions at 60 and 160 K. Therefore, it is not clear at this point whether these new magnetic transitions (Figure 8a) are indicative of an incomplete conversion of the oxide nanoparticles to the fullerene-like sulfides, or perhaps can be related to a strain induced transitions in the *IF* structure. Moreover, a condensed film of oxygen, which is molecular solid at temperatures up to 54.4 K, is known to exhibit similar transitions.[43] In the present measurements, these transitions can also be amplified by oxygen absorption on the large surface area of the nanoparticles.

A transformation from diamagnetic to paramagnetic behavior occurs for the intercalated samples. Furthermore, most of the samples exhibit temperature independent magnetic behavior, as shown in Figure 8b for K-intercalated *IF*-WS$_2$. Such temperature independent magnetic susceptibility, i.e., Pauli paramagnetism, characterizes free electrons.[44] Thus, the transformation from diamagnetic to paramagnetic behavior may be attributed to the charge transfer from the intercalated alkali atoms to the unoccupied *d* bands and to the increasing concentration of the free electron in the host lattice. This is consistent with previous reports regarding alkali metal intercalated 2H-MoS$_2$ and 2H-WS$_2$ materials,[7,15] as well as hydrazine intercalated TiS$_2$.[41] The magnetic susceptibility of the other intercalated samples is estimated to be of $10^{-6} - 10^{-5}$ emu/g.

MSM of the intercalated *IF*-WS$_2$ after storage in the ambient atmosphere for six months reveals that the sample became, once more, diamagnetic (see Figure 8c). These results indicate that deintercalation of the alkali atoms took place. Note, that deintercalated sample has larger negative magnetic moment than the pristine one. This fact indicates that small paramagnetic components exist in the pristine powder, perhaps due to inclusion of some unidentified impurities. During deintercalation, the water of hydration reveals some cleaning effect, taking away not only the intercalated species but also the original impurities. Transport property measurements, which will be discussed below, provide additional evidence for this hypothesis.

5.2. Transport Properties. Figure 9 presents resistivity (ρ) versus temperature measurements for 2H-MoS$_2$ before and after K-intercalation. The pristine sample shows typical semiconductor characteristics (Figure 9a). Sample with lower concentration of the alkali atom (12%) exhibited semiconductor behavior (Figure 9b) too, however, with a reduced resistivity (see insert to Figure 9) when compared with the pristine sample. The sample containing the highest metal concentration (8−18%), as measured by EDS analysis, exhibited normal metallic behavior (Figure 9c). Prolonged storage of the sample with alkali metal concentration of 8−18% in ambient conditions results in deintercalation and, consequently, in reappearance of semiconductor behavior as shown in the Figure 9d.

Resistivity vs temperature measurements were done for all four kinds of pristine materials before and after intercalation. Metallic behavior was not observed in any other sample. However, the room-temperature resistivity and the apparent activation energy (E_A) were affected by the intercalation as summarized in Table 4. All the materials studied show a one to six orders of magnitude reduction in the room-temperature resistivity and a two- to five-fold decrease in E_A, as a result of the intercalation. The values of E_A were calculated from resistivity versus temperature curves using an Arrhenius equation. In general, higher E_A values were obtained for the fullerene-like nanoparticles in comparison to the bulk 2H phase, in both the pristine and intercalated samples. Surprisingly, the room-temperature resistivity of the deintercalated sample became lower than that of the pristine material, while the activation energy increased significantly, approaching a value of the direct band gap (E_g) for 2H-MoS$_2$ single crystal, which varies between 1.74 and 2.27 eV depending on the source of the data.[45] This observation correlates with the magnetic measurements, which

---

(42) *CRC Handbook Chem. Phys*; 69th Edition; CRC Press: Boca Raton, 1988−1989; p E−131 and E−129.
(43) Gregory S. *Phys. Rev. Lett.* **1978**, *40*, 723.
(44) Kittel, C.; *Introduction to Solid State Physics*; John Wiley & Sons: New York, 6th ed., p 396.

(45) Aruchamy, A.; Photoelectrochemistry and Photovoltaics of layered Semiconductors; Kluwer Academic Publishers: Netherlands, p 37.

Table 4. Room Temperature Resistivity and Apparent Activation Energies for Pure and Intercalated 2H−MoS$_2$ and IF-MoS$_2$.

sample	Pristine		K-intercalated[a]		Na-intercalated[a]	
	E_A (eV)	ρ(Ohm*m) at 300 K	E_A (eV)	ρ(Ohm*m) at 300 K	E_A (eV)	ρ(Ohm*m) at 300 K
2H-MoS$_2$	0.55	1.5×10^3	0.45 (12 at. %)	3.3×10^1	0.28 (3 at. %)	7.7
			metal (18 at. %)	6×10^{-3}	0.17 (9 at. %)	4.5
			1.86 (deintercal.)	8×10^2	0.08 (14 at. %)	0.1
IF-MoS$_2$	0.62	7×10^3	0.28 (8 at. %)	2.4×10^1	0.20 (10 at. %)	4.1×10^1

[a] Concentration of the intercalated alkali metal is given in the parentheses.

suggested a cleaning effect of the outgoing water molecules in the deintercalated sample.

In the case of Na-intercalated 2H-MoS$_2$, ρ and E_A were measured for three concentrations of the intercalant (Table 4). The higher the Na at. %, the smaller the resistivity and the activation energy. In the case of K-intercalated 2H-MoS$_2$, increasing the K concentration resulted in a semiconductor to metal transition as shown in Table 4.

If each intercalated atom would contribute a free electron to the conduction band of the host, metallic behavior would be expected for most, if not all, the intercalated samples. Instead, most intercalated samples preserved their semiconductive behavior. Nevertheless, a relatively mild decrease in the resistivity values and activation energy were observed after intercalation. It is important to emphasize that partial oxidation of the exposed samples could occur during the preparation of the contacts in the ambient atmosphere, which may influence their characteristics. Water insertion during or after the metal intercalation could led to neutralization of part of the free carriers, giving a partial explanation for the lack of metallic behavior in the intercalated samples.

Furthermore, metallic behavior requires an intimate contact between the nanoparticles, which is difficult to accomplish with the present method of pellet preparation. Clarifying this point requires further study.

Conclusions

2H-MS$_2$ and IF-MS$_2$ (M = Mo, W) powders have been intercalated by exposure to alkali metal (potassium and sodium) vapor at elevated temperatures. The intercalation did not yield a pure phase. Instead, both the bulk (2H) and the nanopowder (IF) materials led to mixed intercalated/nonintercalated phases. The elastic strain of the closed folded shells of the fullerene-like nanoparticles is believed to be responsible for the incomplete intercalation of these materials, but there is no explanation for the incomplete intercalation of the 2H-phase.

As was demonstrated by XRD, XPS, and EDS/SEM analyses, the intercalation of 2H platelets was found to be more effective (higher at. %) than that of the fullerene-like nanoparticles.

EDS/TEM analysis of the IF-MS$_2$ phase confirms the presence of intercalant atoms in the individual nanoparticles. The integrity of the intercalated IF nanostructures was confirmed by TEM imaging, partial distortions in the outermost layers of the nanoparticles was nonetheless observed. The exposure of the intercalated IF nanoparticles to ambient conditions results in a significant expansion along the c-axis.

A substantial increase ($\approx 3-5$ Å) in the interplanar spacing of the intercalated phase was observed also by XRD analysis, WAXS and by electron diffraction. No changes in the a- and b- lattice constants were observed. Intercalation of one to two water molecules per intercalated atom was suggested as a possible explanation for this large expansion of the c-axis.

The transport and magnetic properties of the intercalated samples changed significantly under intercalation. Heavily (8−18%) K intercalated 2H-MoS$_2$ exhibited a semiconductor to metallic transition. A transition from diamagnetic to paramagnetic behavior as well as a decrease in room-temperature resistivity and activation energy values were observed for all the intercalated phases.

Deintercalation of the hydrated alkali atoms and restacking of the MS$_2$ layers was observed by WAXS, XRD, and TEM analyses in all the samples after prolonged exposure to the atmosphere, which was confirmed by recovering of the pristine host compound properties.

Field-induced deintercalation of the particles was also observed, above a threshold voltage of the flood-gun, by XPS technique.

Reversible intercalation of alkali atoms into the host IF lattice was demonstrated in this study. However, it is clear that the loading of the nanoparticles by the present method is rather limited. For any foreseeable application, loading of the nanoparticles with metal atoms would have to be increased. It is clear that the large surface area of the IF nanopowder could be advantageous for electrode material, provided it would contain sufficient reactive sites for the intercalation/deintercalation process. However, the inert van der Waals surface of the closed nanoparticles presents a diffusion barrier for the intercalation process. Once the alkali atoms are intercalated into the outer closed layers of the IF nanoparticle, their large affinity toward the solvent molecules (water in the present case) leads to the formation of defects in the host lattice and, consequently, to a very rapid intercalation of water molecules between the layers. In the partially ruptured IF layers, the process could be easily reversed as indicated by the present experiments. Therefore, it is believed that a few cycles of this kind or perhaps a few cycles of intercalation/deintercalation under electrochemical control would make this phase ideally suitable for reversible alkali metal intercalation. The ramifications of this process can be of substantial importance, e.g. to the battery industry.

Acknowledgment. We are grateful to Dr. R. Rosentsveig and A. Margolin for assistance with the synthesis of the IF-MoS$_2$ and IF-WS$_2$ materials; to Dr. C. Lévy-Clément, CNRS-Thiais, Dr. I. B. Zubavichus, Nesmeyanov Institute of Organoelement Compounds, Moscow and Prof. S. Shilstein, Weizmann Institute of Science for enlightening discussions; to Dr. Y. Rosenberg, Tel-Aviv University for assistance in XRD measurements. This work was supported by the Minerva Foudation (Munich) and Alfried Krupp von Bohlen und Halbach Steiftung.

JA012060Q

contained two to four nano-octahedra and some amorphous material. The intensity of the S K peak was significantly reduced in the spectrum of the nano-octahedra: the S/Mo ratio of 1.3 ± 0.1 corresponds to a sulfur deficiency of approximately 35% with respect to the composition of the quasispherical fullerene-like nanoparticles. Several tests were done to confirm that no beam damage took place during the EELS analysis. The measured sulfur deficiency may also reflect a low S/Mo ratio in the amorphous material between the nano-octahedra.

Quantitative analysis of the low-energy regions of the spectra was also undertaken. The plasmonic part of the spectrum of the nano-octahedra was compared with those of the larger quasispherical fullerene-like nanoparticles and the 2H-MoS_2 plates. For the fullerene-like MoS_2 nanoparticles, the statistically averaged peak occurred at (24.6 ± 0.3) eV and its full width at half maximum was 9.8 eV; for the nano-octahedra, the peak occurred at (23.7 ± 0.4) eV and had a width of 16.9 eV. The broadening of the plasmon peak in the spectrum of the nano-octahedra can be ascribed to the large contribution from the surface plasmon near 16 eV.[19] This contribution may also be the reason for the shift of the plasmon peak to lower energy. The overall similarity between the two spectra is an indication of the close relationship between the nano-octahedra and the 2H-MoS_2 plates.

An example of an electronic density of states (DOS) curve calculated for an octahedral Mo_xS_{2x-12} nanostructure is shown in Figure 4. The DOS profiles of the fullerenes are quite similar to those of semiconducting MoS_2 nanotubes and bulk 2H-MoS_2.[4] Although the general features of the electronic structures are the same in all these cases, those of the hollow fullerenes exhibit metallic character. The gap between the highest occupied molecular orbital (HOMO) and the lowest unoccupied molecular orbital (LUMO) does not exceed a few hundredths of an electronvolt, irrespective of the size of the nano-octahedron. The valence band (-7.5 to 1.5 eV) has mixed Mo 4d and S 3p character. The states near the HOMO (Fermi energy) and the LUMO of the octahedral Mo_xS_{2x-12} fullerenes have mainly Mo 4d character.

A Mulliken charge-distribution analysis reveals a charge transfer from the molybdenum to the sulfur atoms. The charges in an MoS_2 monolayer or nanotube are -0.453 and $+0.906$ for the sulfur and molybdenum atoms, respectively.[4] For an octahedral Mo_xS_{2x-12} fullerene (for example, $Mo_{576}S_{1140}$) average charges of -0.41 and -0.47 are calculated for the internal and external sulfur atoms, respectively, and a charge of $+0.91$ is calculated for the molybdenum atoms. Similar differences between the charges of the internal and external sulfur atoms were found for MoS_2 nanotubes.[4]

In summary, atomic models of molybdenum sulfide nanoparticles with fullerene-like structures were constructed. For the first time, their stabilities and electronic properties were investigated as a function of particle size using the DFTB method. Our calculations and MD simulations indicate that stoichiometric single-walled MoS_2 fullerenes with octahedral shapes and sizes of a few hundred atoms are unstable. This instability is attributed to the high strain energy at the corners of the MoS_2 nano-octahedra. Furthermore, in good agreement with the experimental data, the model calculations show that multiwalled nano-octahedra are stable in a limited size range of 10^4–10^5 atoms. Larger particles are converted into multiwalled MoS_2 nanoparticles with quasispherical shapes.

The electronic properties of the molybdenum sulfide nano-octahedra are found to be radically different those of the bulk solid. Nanoplatelets, nanotubes, and quasispherical nanoparticles of MoS_2 are semiconductors. In contrast, irrespective of their size and exact stoichiometry, all hollow octahedral molybdenum sulfide fullerenes exhibit a nearly vanishing gap between the HOMO and the LUMO, which are mainly of Mo 4d character.

The results presented herein should stimulate further experimental investigations, which will lead to the elucidation of the structure and the physical properties of the inorganic fullerenes and related nanostructures of MoS_2 and other layered compounds.

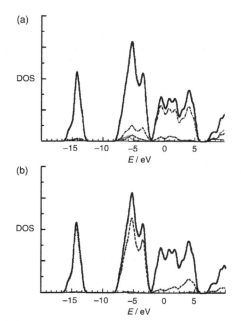

Figure 4. Molybdenum (a) and sulfur (b) DOS curves for an $Mo_{100}S_{188}$ nano-octahedron (calculated using the DFTB method). d states –·–·, p states ----, s states ·····; 0 eV corresponds to the HOMO energy (Fermi energy).

Experimental Section

All calculations were performed using the DFTB method[20] with full geometry optimization. MD simulations (constant number, volume, and temperature (NVT) ensemble) were performed for the optimized structures. The energies in Figure 3 are given in Hartree units (1 H = 27.2114 eV).

MoS_2 nano-octahedra: MoS_2 powder (Sigma Aldrich, 99.5%) was pressed into a pellet (diameter of 17 mm). The samples were heated at 450–700°C in a quartz reactor. Laser ablation was conducted using pulses from a mildly focused frequency-doubled Nd:YAG laser (532 nm, 10 Hz, 8 ns, ca. 60 mJ per pulse, duration 20 min). The beam was scanned continuously across the pellet surface. The generated soot was flushed back downstream by flowing argon/helium gas

(760 Torr, 200 cm³ min⁻¹) and was collected on a quartz substrate outside the oven.

The collected powder was sonicated in ethanol, placed on a carbon/collodion-coated copper grid, and analyzed by TEM (Philips CM-120, 120 kV). Furthermore, HRTEM (FEI Tecnai F-30, 300 kV) with an imaging filter (Gatan GIF) was used for EELS. Energy dispersive X-ray spectroscopy (EDS; EDAX Phoenix) was also used.

Received: May 29, 2006
Revised: August 16, 2006
Published online: December 12, 2006

Keywords: density functional calculations · electronic structure · fullerenes · molecular dynamics · nanoparticles

---

[1] R. Tenne, L. Margulis, M. Genut, G. Hodes, *Nature* **1992**, *360*, 444–446.
[2] L. Margulis, G. Salitra, M. Talianker, R. Tenne, *Nature* **1993**, *365*, 113–114.
[3] Y. Feldman, E. Wasserman, D. J. Srolovitz, R. Tenne, *Science* **1995**, *267*, 222–225.
[4] G. Seifert, H. Terrones, M. Terrones, G. Jungnickel, T. Frauenheim, *Phys. Rev. Lett.* **2000**, *85*, 146–149.
[5] G. Seifert, T. Köhler, R. Tenne, *J. Phys. Chem. B* **2002**, *106*, 2497–2501.
[6] I. Kaplan-Ashiri, S. R. Cohen, K. Gartsman, R. Rosentsveig, G. Seifert, R. Tenne, *J. Mater. Res.* **2004**, *19*, 454–459.
[7] I. Kaplan-Ashiri, S. R. Cohen, K. Gartsman, V. Ivanovskaya, T. Heine, G. Seifert, I. Wiesel, H. D. Wagner, R. Tenne, *Proc. Natl. Acad. Sci. USA* **2006**, *103*, 523–528.
[8] L. Rapoport, N. Fleischer, R. Tenne, *J. Mater. Chem.* **2005**, *15*, 1782–1788.
[9] R. Tenne, *Adv. Mater.* **1995**, *7*, 965–972, 989–995.
[10] L. Margulis, S. Iijima, R. Tenne, *Microsc. Microanal. Microstruct.* **1996**, *7*, 87–89.
[11] P. A. Parilla, A. C. Dillon, K. M. Jones, G. Riker, D. L. Schulz, D. S. Ginley, M. J. Heben, *Nature* **1999**, *397*, 114.
[12] P. A. Parilla, A. C. Dillon, B. A. Parkinson, K. M. Jones, J. Alleman, G. Riker, D. S. Ginley, M. J. Heben, *J. Phys. Chem. B* **2004**, *108*, 6197–6207.
[13] J. A. Ascencio, M. Perez-Alvarez, L. M. Molina, P. Santiago, M. José-Yacaman, *Surf. Sci.* **2003**, *526*, 243–247.
[14] A. N. Enyashin, V. V. Ivanovskaya, Yu. N. Makurin, A. L. Ivanovskii, *Inorg. Mater.* **2004**, *40*, 395–399.
[15] A. N. Enyashin, A. L. Ivanovskii, *Russ. J. Phys. Chem.* **2005**, *79*, 1081–1086.
[16] A. N. Enyashin, G. Seifert, *Phys. Status Solidi B* **2005**, *242*, 1361–1370.
[17] J. D. Fuhr, J. O. Sofo, A. Saul, *Phys. Rev. B* **1999**, *60*, 8343–8347.
[18] D. J. Srolovitz, S. A. Safran, M. Homyonfer, R. Tenne, *Phys. Rev. Lett.* **1995**, *74*, 1779–1782.
[19] H. Cohen, T. Maniv, R. Tenne, Y. Rosenfeld Hacohen, O. Stephan, C. Colliex, *Phys. Rev. Lett.* **1998**, *80*, 782–785.
[20] D. Porezag, T. Frauenheim, T. Köhler, G. Seifert, R. Kashner, *Phys. Rev. B* **1995**, *51*, 12947–12957.

Toward Atomic-Scale Bright-Field Electron Tomography for the Study of Fullerene-Like Nanostructures

Maya Bar Sadan,[†] Lothar Houben,*,[‡] Sharon G. Wolf,[§] Andrey Enyashin,[∥] Gotthard Seifert,[∥] Reshef Tenne,[†] and Knut Urban[‡]

Materials and Interfaces Department, Weizmann Institute of Science, Rehovot 76100, Israel, Institute of Solid State Research and Ernst-Ruska Centre for Microscopy and Spectroscopy with Electrons, Research Centre Jülich GmbH, 52425 Jülich, Germany, Electron Microscopy Unit, Weizmann Institute of Science, Rehovot 76100, Israel, and Physikalische Chemie, Technische Universität Dresden, 01062 Dresden, Germany

Received December 3, 2007; Revised Manuscript Received January 16, 2008

ABSTRACT

We present the advancement of electron tomography for three-dimensional structure reconstruction of fullerene-like particles toward atomic-scale resolution. The three-dimensional reconstruction of nested molybdenum disulfide nanooctahedra is achieved by the combination of low voltage operation of the electron microscope with aberration-corrected phase contrast imaging. The method enables the study of defects and irregularities in the three-dimensional structure of individual fullerene-like particles on the scale of 2–3 Å. Control over shape, size, and atomic architecture is a key issue in synthesis and design of functional nanoparticles. Transmission electron microscopy (TEM) is the primary technique to characterize materials down to the atomic level, albeit the images are two-dimensional projections of the studied objects. Recent advancements in aberration-corrected TEM have demonstrated single atom sensitivity for light elements at subångström resolution. Yet, the resolution of tomographic schemes for three-dimensional structure reconstruction has not surpassed 1 nm³, preventing it from becoming a powerful tool for characterization in the physical sciences on the atomic scale. Here we demonstrate that negative spherical aberration imaging at low acceleration voltage enables tomography down to the atomic scale at reduced radiation damage. First experimental data on the three-dimensional reconstruction of nested molybdenum disulfide nanooctahedra is presented. The method is applicable to the analysis of the atomic architecture of a wide range of nanostructures where strong electron channeling is absent, in particular to carbon fullerenes and inorganic fullerenes.

Tomographic structure reconstruction is a well-established technique in transmission electron microscopy (TEM) in the field of biological sciences.[1–5] In contrast, there has been reluctance to use electron tomography in TEM as a basic tool for the analysis of solid-state structures in the physical sciences. Valuable shape and size information with a resolution of about 1 nm³ is attained through recent progress in high-angle annular dark-field and energy-filtered tomography.[6–8] However, crucial information on molecular architecture or atomic coordination in a tomogram of individual nanoparticles demands even higher resolution on the scale of about 2 Å. Knowledge of the three-dimensional structure and composition on the atomic scale holds the key to understanding the unique physical properties of nanomaterials compared with their bulk counterparts. It is therefore one of the primary goals of electron tomography in material science to improve the resolution.

While bright-field high-resolution TEM offers the desired resolution, there are hitherto two main obstacles for atomic resolution bright-field tomography. First a resolution of around 2 Å could so far be achieved only at higher acceleration voltages, where severe radiation damage inhibits the recording of a sufficient number of tilt images of a single object. Second, bright-field images of crystalline material are subject to nonlinear dynamical diffraction and interference of the diffracted electron upon image recording. This implies that the "projection requirement", i.e., the requirement of the recorded signal to be related to a physical property in a monotonous manner, is usually not fulfilled.

Here we propose the application of low-voltage aberration-corrected bright-field electron tomography to overcome these obstacles to obtain three-dimensional (3D) reconstructions with atomic resolution of architectures without translatory periodicity and strong electron channelling effects. Therefore this technique is particularly applicable to fullerene structures, e.g., of inorganic layered compounds.[9–11] Here we will focus

* Corresponding author: l.houben@fz-juelich.de.
[†] Materials and Interfaces Department, Weizmann Institute of Science.
[‡] Institute of Solid State Research and Ernst-Ruska Centre for Microscopy and Spectroscopy with Electrons, Research Centre Jülich GmbH.
[§] Electron Microscopy Unit, Weizmann Institute of Science.
[∥] Physikalische Chemie, Technische Universität Dresden.

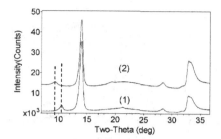

Figure 7. XRD patterns of the freshly intercalated powder (1) and of the same powder after three months of storage (2). The (002′) peaks are marked by dashed lines.

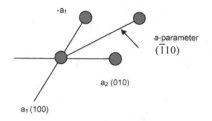

Figure 8. In-plane crystal directions in 2H-WS$_2$; solid circles represent either metal or chalcogen atoms.

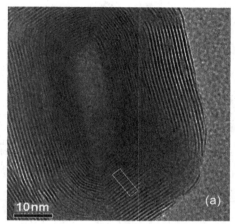

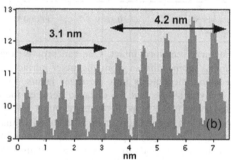

Figure 9. (a) TEM image of a typical rubidium intercalated IF particle and (b) the intensity profile of the framed layers at the interface between the pristine (inner) and the intercalated (outer) lattices. The difference between the outermost WS$_2$ layers (Rb intercalated) and the unmodified inner layers is clear.

a-axis value of the intercalated phase in the Rb$_x$IF_a was found to be 3.25 Å rather than 3.27 Å for the regularly intercalated IF-WS$_2$ phase (see Table 4).

Furthermore, Figure 7 compares two XRD patterns of another rubidium intercalated IF powder. One of the patterns was obtained on a freshly prepared powder, while the other was obtained on that sample after 3 months of storage time (in the glovebox) during which the material had absorbed some water. It was found that in this case the c-axis of the intercalated phase had expanded from 16.71 Å up to 18.62 Å, whereas the a-axis had shrank to 3.24 Å compared with 3.27 Å for the freshly intercalated phase. It can be seen that both depicted cases (rubidium intercalated IF powders after 3 months of storage and when the metal grains and IF nanoparticles had mixed) showed the same tendency; that is, when the (002′) line of the intercalated phase moved to lower angles, i.e., to larger interlayer spacing, the (100′) and (110′) peaks, characterizing the a-parameter, shifted to higher angles from their initial values, respectively, Tables 2 and 4. Note, however, that in both cases the a-parameter is increased compared with the pristine IF-WS$_2$ material (3.16 Å). Increase in the a-parameter upon intercalation can be explained considering the band structure of WS$_2$. The top of the valence band of WS$_2$ contains d$_{z^2}$[21] orbitals and chalcogen p states while the conduction band is derived from the non bonding d$_{xy}$ and d$_{x^2-y^2}$ orbitals.[9] Previous studies[22] showed that the pristine IF material was a p-type semiconductor. It might be assumed that upon intercalation electrons are transferred from the rubidium atoms to the empty states in the conduction band. Consequently, the electron density increases within the plane of the metal atoms and between coplanar S atoms and this phenomenon leads to the repulsion between adjacent sulfur or tungsten atoms (Figure 8).[4] The increased electron density in the plane leads to a repulsion between tungsten atoms and between sulfur atoms as well.[4] The experimental results presented above show that while the interlayer distances grew by 1.91 Å due to the water absorption, the expanded a-parameter was reduced only by 0.03 Å compared to the freshly intercalated sample, Tables 2 and 4. Probably some charge transfer to the water molecules occurred leading to a reduced charge density at the W atom and consequently reduction in the a-parameter.

Another interesting observation was the presence of the unidentified peaks in the pattern of the rubidium intercalated IF powder (Figure 6b). These lines were observed in the two cases where the percentage of the intercalated phase was the highest. Further experiments are needed for a complete assignment of these peaks.

TEM and Electron Energy Loss Spectroscopy (EELS)/ TEM Analysis. High Resolution TEM analysis was done for the rubidium intercalated phase, since this metal gives the greatest layer expansion and the Z-contrast between the host and the guest is the highest. Furthermore, the Rb-intercalated IF phase was not studied before. As can be seen in Figure 9 the outermost layers of the closed IF nanostructures expanded significantly suggesting that they underwent intercalation, whereas the inner layers of the closed IF

---

(21) Ito, T.; Iwami, M.; Hiraki, A. *J. Phys. Soc. Jpn.* **1981**, *50*, 106.
(22) Kopnov, F.; Yoffe, A.; Leitus, G.; Tenne, R. *Phys. Status Solidi B* **2006**, *243*, 1229.

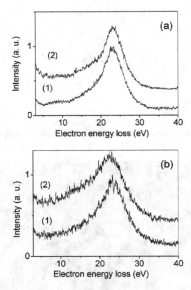

Figure 10. Typical electron energy loss spectra of (a) a pristine IF nanoparticle, (1) the whole particle and (2) the edge of the nanoparticle, and (b) a Rb intercalated IF particle, (1) the whole particle and (2) the edge of the particle.

Table 5. Width at Half Maximum of the EELS Spectra

type of the material	fwhm of the whole particle	fwhm of the edge of the particle
IF pristine	6.5 (±0.3) eV	9.0 (±0.5) eV
Rb intercalated IF particle	7 (±0.3) eV	11 (±0.5) eV

nanostructures remained intact. One can propose that the alkali metal atoms diffuse from the outer surface of the particle along the defects into the inner layers. Intensity profile analysis shows that the distance between the outer expanded layers is 8.4 Å and between the intact inner layers is 6.2 Å, which is in good agreement with the XRD analysis. Obviously, the interlayer expansion due to the intercalation of the closed layers leads to a significant accumulation of strain in the IF nanoparticle. The deeper the penetration of the alkali atom into the core of the IF nanoparticle, the larger the accumulated strain. This strain acts against the chemical driving force for the intercalation until an equilibrium state is attained. Unfortunately, EDS and EELS/TEM analyses of rubidium intercalated nanoparticles are practically impossible, due to the overlap between the rubidium and tungsten lines.

The low loss EELS region of the rubidium intercalated and pristine IF particles showed slight differences (Figure 10). The spectra were taken once when the electron beam covered the entire nanoparticle and also when only the edge of the nanoparticle was exposed to the focused electron beam. It can be seen that for the pristine nanoparticles both spectra (taken from the entire nanoparticle and its edge); (Figure 10a) consist of a broad hump between 6 and 11 eV and a main peak at 23.2 eV. These features are typical of bulk WS_2.[23] Nonetheless, the EELS spectrum from the nanoparticle edge is somewhat broader and asymmetric (Table 5) due to enhanced contribution of the surface plasmons.[24] The

(23) Bell, M. G.; Liang, W. Y. *Adv. Phys.* **1976**, *25*, 53.

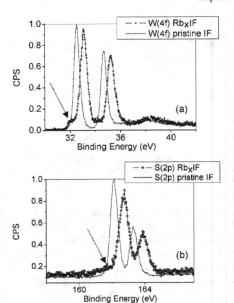

Figure 11. XPS windows of rubidium intercalated and pristine IF powder: (a) W(4f), (b) S(2p). Arrow points to the shoulder in each spectrum.

Table 6. Energy Band Shifts $E_{intercalated} - E_{pristine}$ of the Rb Intercalated Compared with the Pristine IF-WS_2 Phase

band	W(4f)	S(2p)	valence band (VB) shift	VB threshold shift
energy shift (mV)	572 (±50)	600 (±50)	600 (±50)	200 (±50)

prominent peak at 23.2 eV is associated with the bulk plasmon excitation, involving the entire valence electron gas of the medium. The broad hump (centered around 8.5 eV) might be connected to the plasmon resonance which hybridizes with interband electron transitions.[23] The two corresponding spectra of the Rb intercalated particle, Figure 10b, are rather broad (Table 5) as compared to the spectra of the pristine particle, and the hump around 8.5 eV is washed out. The larger width of the plasmonic band in the intercalated particle may reflect lower homogeneity and may also be influenced by the rubidium single electron transitions associated with[25] the Rb($4p_{3/2}$) level at 15.2 eV and Rb ($4p_{1/2}$) at 16.1 eV.

XPS Characterization. Rubidium intercalated IF powder was characterized by X-ray photoelectron spectroscopy. The XPS measurements revealed that the elemental binding energies and the valence band of the intercalated sample shifted to higher energies by the same amount (600 mV) as compared to the pristine sample, Table 6. Elimination of charging was obtained by systematically varying the charging conditions from positive to negative values. These line shifts showed the same tendency as in our pervious work on intercalation of IF nanoparticles with sodium and potassium.[26] In contrast, the top of the valence band of the intercalated powder shifted by 200 mV only, Table 6, indicating the introduction of new (impurity) states into the

(24) Egerton, R. F. *Electron energy-loss spectroscopy in the electron microscope*, 2nd ed.; Plenum Press: New York, 1996.
(25) Ebbinghaus, G.; Simon, A. *Chem. Phys.* **1979**, *43*, 117.

forbidden gap. Figure 11a,b shows the W(4f) and S(2p) spectra of the intercalated and pristine powders. The shifts in binding energies of the intercalated IF powder to higher values are interpreted as an upward shift in the Fermi level, E_F; that is, the intercalated material becomes more electron rich than the pristine one. Note that both of the W(4f) and S(2p) bands of the intercalated material demonstrate a low binding energy shoulder (Figure 11a,b, arrows). These shoulders do not arise from differential charging, as proven by a systematic variation of the charging conditions (not shown). Thus, they are associated with a charge transfer (from the intercalant, Rb), where this extra charge remains relatively localized in the host lattice, for example, near defects like corners[27] and so forth.

On the basis of the XPS data only, it is difficult to estimate the atomic percentage of the intercalated rubidium. Clearly, part of the Rb signal corresponds to oxidized metal phase covering the surface of the nanoparticles. This fact was concluded from in situ chemically resolved electrical measurements (CREM).[28] By switching on the electron flood gun (eFG) to get a sample current of 60 nA, the W(4f) and S(2p) lines shifted by 50 mV ($\pm$0.15 mV) only, while the Rb(3d) line shifted by 310 mV ($\pm$0.15 mV). The CREM data (including additional eFG conditions, not shown) indicated that a major part of the detected Rb was situated externally to the nanoparticles, and yet was in direct contact with them (e.g., as an external coating). This Rb component was oxidized and was too dominant to allow the direct quntitative estimate of the intercalated Rb. In fact, the indirect information mentioned above, that is, the upward shift of the Fermi energy, was used as the evidence for the existence of the charge transfer from the guest (Rb atom) to the host WS_2 lattice. This charge transfer is believed to be the driving force for the intercalation process.[1]

Conclusions

IF-WS_2 nanoparticles and 2H-WS_2 (bulk) were subjected to the intercalation process with sodium, potassium, and rubidium. Special care was exerted to keep the intercalated phases in a moisture free environment during the study. The XRD analysis under the inert conditions showed that the intercalated material was inhomogeneous and was composed of several phases. The product was made up from intercalated and nonintercalated phases. Many of the nanoparticles were intercalated only at their outermost layers with the core remaining unchanged. Additionally Na_2S was formed in the course of the reaction of IF-WS_2 and sodium, and some potassium sulfide phases were also found in the intercalated nanoparticles. The van der Waals gap expansion of the intercalated particles depended on the radius of the alkali atom: the greatest interlayer expansion was obtained for the rubidium intercalation, the least for the intercalation with sodium atoms. The a-axis of the intercalated phase was found to expand as well. The fraction of the intercalated phases in the reacted (IF and 2H) powders depended on the specific alkali metal, with highest percentage of the intercalated phase achieved for rubidium intercalation in both kinds of powders. However, the amount of the intercalated IF phase was higher than the quantity of the intercalated 2H-WS_2 phase with the same metal under the same conditions. This observation could be associated with the difference in the surface to volume ratio of the two kinds of particles, which is greater by order of magnitude for the IF nanoparticles. The EELS analysis showed a broader plasmon line in the case of the rubidium intercalated particles. The XPS measurements of the rubidium intercalated IF-WS_2 powder revealed an upward shift in the Fermi level, E_F, corresponding to a guest-to-host electron transfer, and additionally indicated the existence of external coating on the nanoparticles, consisting of oxidized rubidium.

Acknowledgment. R.T. is the holder of the Drake family Chair in Nanotechnology and the Director of the Helen and Martin Kimmel Center for Nanoscale Science. This work was supported by "NanoMaterials, Ltd.", the GMJ Schmidt Minerva Center, the Harold Perlman Foundation, and the Israel Science Foundation.

CM800020P

---

(26) Feldman, Y.; Zak, A.; Tenne, R.; Cohen, H. *J. Vac. Sci. Technol., A* **2003**, *21*, 1752.
(27) Panich, A.; Kopnov, F.; Tenne, R. *J. Nanosci. Nanotechnol.* **2006**, *6*, 1.
(28) Cohen, H. *Appl. Phys. Lett.* **2004**, *85*, 1271.

our model that the oscillations in $\tau(\phi)$ beyond ϕ_{crit} come from the undulations in $E_{vdW}(\phi)$. The theoretical period of the oscillations derived from our model depends on diameter and varies from 2.3° for the 16 nm diameter nanotube to 1.2° for the 32 nm diameter nanotube. The observed torque oscillations have periods of $2 \pm 1°$, which are consistent with our model within experimental error.

The diameter dependence of ϕ_{crit} obtained from the model by repeating the energy calculations for different diameters is plotted in Fig. 2 by stars. The values of ϕ_{crit} from the model and from the data show similar diameter dependences, although the absolute values differ by nearly a factor of 2. This difference could arise from the uncertainty in the determination of the shear modulus, and also from the approximate evaluation of the interlayer vdW energy. However, the semiquantitative agreement with the experimental values from such a simple calculation strongly supports our model.

Overall, our simple energy-based model gives a qualitatively accurate description of the observed behavior. It (i) shows that there is a crossover from stick to slip motion, (ii) reproduces the diameter dependence of ϕ_{crit}, and (iii) shows that above ϕ_{crit} the $\tau(\phi)$ oscillates. By reproducing all the features of the experimental data, this model indicates that the stick-slip behavior comes from the torsional energy cost becoming larger than the vdW energy cost upon twisting.

In summary, we observed atomic-scale torsional stick-slip behavior in WS_2 NTs. The behavior consists of a stick regime where all the nanotube walls twist together, and a slip regime where the outer wall slips over the inner walls. A simple theoretical model allowed us to determine the origin of this torsional stick-slip behavior to be the competition between the vdW and torsional energies. In the slip regime, the corrugation in the vdW energy gives rise to secondary stick-slip oscillations. The observed stick-slip behavior and the oscillations in the slip regime come about due to the commensurate atomic arrangement of the WS_2 layers in a WS_2 NT and their large interlayer corrugation energy combined with their relatively small in-plane shear stiffness.

We thank R. Rosentsveig for the WS_2 nanotube synthesis, M. Bar-Sadan, J. Klein, and H. D. Wagner for helpful discussions, Y. Wang for the DFTB calculations, and A. Yoffe for assistance with the clean room. This research was supported by the Israel Science Foundation, German-Israeli foundation, Minerva Stiftung, Kimmel Center for Nanoscale Science, Israeli Ministry of Defense, Djanogly and Alhadeff and Perlman foundations. The EM studies were carried out at the Moskovitz Center for Nano and Bio Imaging. E. J. received support from the Victor Erlich Career Development Chair program. K. S. N. acknowledges Feinberg Graduate School for support.

---

*nagapriya.kavoori@weizmann.ac.il
+ernesto.joselevich@weizmann.ac.il

[1] A. N. Cleland, *Foundations of Nanomechanics* (Springer, Berlin, 2003).
[2] B. Bhushan, *Nanotribology and Nanomechanics: An Introduction* (Springer, Berlin, 2005).
[3] C. H. Scholz, Nature (London) **391**, 37 (1998).
[4] J. Klein, Phys. Rev. Lett. **98**, 056101 (2007).
[5] *Fundamentals of Friction; Macroscopic and Microscopic Processes*, edited by I. L. Singer and H. Polland, NATO ASI Series (Kluwer, Dordrecht, 1991).
[6] Y. A. Khulief, F. A. Al-Sulaiman, and S. Bashmal, J. Sound Vib. **299**, 540 (2007).
[7] A. Kis *et al.*, Phys. Rev. Lett. **97**, 025501 (2006).
[8] W. Guo, W. Zhong, Y. Dai, and S. Li, Phys. Rev. B **72**, 075409 (2005).
[9] A. Barreiro *et al.*, Science **320**, 775 (2008).
[10] P. A. Williams *et al.*, Phys. Rev. Lett. **89**, 255502 (2002).
[11] T. Cohen-Karni *et al.*, Nature Nanotech. **1**, 36 (2006).
[12] K. S. Nagapriya *et al.*, arXiv:0803.4426v1.
[13] R. Tenne, L. Margulis, M. Genut, and G. Hodes, Nature (London) **360**, 444 (1992).
[14] R. Tenne, Nature Nanotech. **1**, 103 (2006).
[15] J. D. Fuhr, J. O. Sofo, and A. Saúl, Phys. Rev. B **60**, 8343 (1999).
[16] I. Kaplan-Ashiri *et al.*, J. Phys. Chem. C **111**, 8432 (2007).
[17] B. Bourlon *et al.*, Nano Lett. **4**, 709 (2004).
[18] S. B. Trickey, F. Müller-Plathe, G. H. F. Diercksen, and J. C. Boettger, Phys. Rev. B **45**, 4460 (1992).
[19] M. Bar-Sadan *et al.*, Proc. Natl. Acad. Sci. U.S.A. **105**, 15 643 (2008).
[20] K. Hirahara *et al.*, Phys. Rev. B **73**, 195420 (2006).
[21] I. Kaplan-Ashiri *et al.*, J. Mater. Res. **19**, 454 (2004).
[22] I. Kaplan-Ashiri *et al.*, Proc. Natl. Acad. Sci. U.S.A. **103**, 523 (2006).
[23] R. Rosentsveig *et al.*, Chem. Mater. **14**, 471 (2002).
[24] See EPAPS Document No. E-PRLTAO-101-040846 for additional torque-torsion curves, additional torsion-time curves, a movie of pedal rotation due to charging, and SEM images of twisted nanotubes showing the absence of buckling. For more information on EPAPS, see http://www.aip.org/pubservs/epaps.html.
[25] M.-F. Yu, T. Kowalewski, and R. S. Ruoff, Phys. Rev. Lett. **86**, 87 (2001).
[26] Y. Shibutani and S. Ogata, Model. Simul. Mater. Sci. Eng. **12**, 599 (2004).
[27] B. Yakobson and P. Avouris, *Carbon Nanotubes: Synthesis, Structure, Properties, and Applications*, Topics in Applied Physics (Springer-Verlag, Berlin, 2001), Vol. 80, pp. 287–327.
[28] E. Ertekin and D. C. Chrzan, Phys. Rev. B **72**, 045425 (2005).

Inorganic Fullerenes

DOI: 10.1002/anie.201006719

MoS$_2$ Hybrid Nanostructures: From Octahedral to Quasi-Spherical Shells within Individual Nanoparticles**

Ana Albu-Yaron, Moshe Levy, Reshef Tenne,* Ronit Popovitz-Biro, Marc Weidenbach, Maya Bar-Sadan, Lothar Houben, Andrey N. Enyashin, Gotthard Seifert, Daniel Feuermann, Eugene A. Katz, and Jeffrey M. Gordon*

MoS$_2$, a layered compound with tribological and catalytic applications, is known to form a range of hollow closed nanostructures[1–5] and nanoparticles, including graphene-like structures.[6] These have been demonstrated experimentally through high-temperature synthesis and pulsed laser ablation (PLA), and theoretically with quantum chemical calculations. The smallest allowed structures are nanooctahedra of 3 to 8 nm size. Nicknamed the "true inorganic fullerene" in analogy to carbon fullerenes,[4] they differ from larger multi-walled MoS$_2$ fullerene-like nanoparticles both in their morphology and predicted electronic properties. The larger fullerene-like particles are quasi-spherical (polyhedral) or nanotubular, typically with diameters of 20 to 150 nm.[3] Above a few hundred nm in size, these nanoparticles transform into 2H-MoS$_2$ platelets. Fullerene-like particles have been recognized as superior solid lubricants[7] with numerous commercial applications, and MoS$_2$ nanooctahedra may have catalytic applications. Understanding the fundamental commonality of these two morphologies might prove essential in the development of new materials.

[*] Dr. A. Albu-Yaron, Prof. Dr. M. Levy, Prof. Dr. R. Tenne
Department of Materials and Interfaces
Weizmann Institute of Science, Rehovot 76100 (Israel)
E-mail: reshef.tenne@weizmann.ac.il

Prof. Dr. D. Feuermann, Dr. E. A. Katz, Prof. Dr. J. M. Gordon
Department of Solar Energy and Environmental Physics, Jacob Blaustein Institutes for Desert Research, Ben-Gurion University of the Negev, Sede Boqer Campus 84990 (Israel)
E-mail: jeff@bgu.ac.il

Dr. R. Popovitz-Biro
Electron Microscopy Unit
Weizmann Institute of Science, Rehovot 76100 (Israel)

M. Weidenbach, Dr. M. Bar-Sadan, Dr. L. Houben
Institute of Solid State Research and Ernst Ruska Centre for Microscopy and Spectroscopy with Electrons, Forschungszentrum Jülich GmbH (Germany)

Dr. A. N. Enyashin, Prof. Dr. G. Seifert
Physikalische Chemie, Technische Universität Dresden (Germany)

[**] R.T. gratefully acknowledges the support of ERC project INTIF 226639, the Israel Science Foundation, the Harold Perlman Foundation, and the Irving and Cherna Moskowitz Center for Nano and Bio-Nano Imaging. R.T. is the Drake Family Chair of Nanotechnology and director of the Helen and Martin Kimmel Center for Nanoscale Science.

Re-use of this article is permitted in accordance with the Terms and Conditions set out at http://onlinelibrary.wiley.com/journal/10.1002/(ISSN)1521-3773/homepage/2002_onlineopen.html

The research on hollow MoS$_2$ nanostructures of minimal size (< 10 nm in diameter) was initiated in 1993 upon the first independent proposal of the formation of nanooctahedra of MoS$_2$[3,5] (and BN)[8] with six rhombi in their corners. In 1999, it was demonstrated[4] that two- to four-walled MoS$_2$ nanooctahedra, 3–5 nm in size and up to ca. 10^4 atoms, could be obtained by PLA. Similar results were subsequently reported in Ref. [1,2] as illustrated in Figure 1a. Recent studies of high energy density methods such as laser ablation and arc–discharge[9–13] resulted in small structures with only a limited number of atoms: Mo–S clusters or double-walled nanooctahedra.

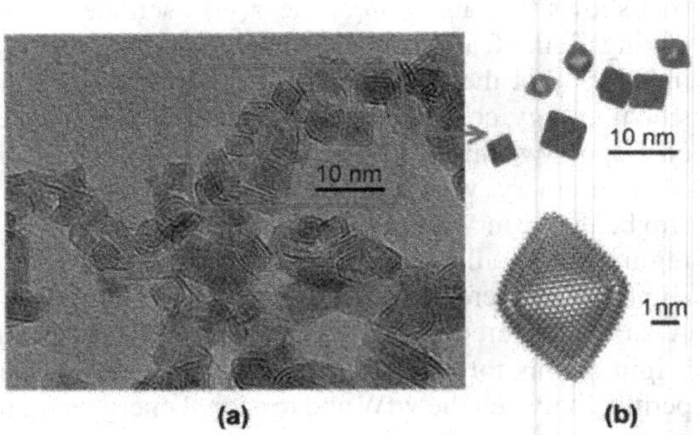

Figure 1. a) Transmission electron microscope (TEM) image that includes MoS$_2$ nanooctahedra generated from MoS$_2$ powder by PLA as reported in Ref. [1,2]. Note the large number of nanooctahedra with two or three layers and < 8 nm in size. b) Ball-and-stick models showing assorted projections of the nanooctahedra in (a), where Mo atoms are red and S atoms are yellow.

On the other hand, theoretical modeling[1,2] suggested that the stability limit for nanooctahedra should be almost an order of magnitude larger than the largest structures produced by PLA (see Figure 3 of Ref. [1]). Therefore, nanooctahedra with more than five layers were not considered in modeling procedures. A principal insight from the PLA studies was that the low density of the Mo–S vapor, coupled with the rapid temperature quench, prevented the nanooctahedra from growing to sizes large enough to realize their stability limit. It was concluded that a two-step process of PLA at 2000°C, followed by in situ annealing at 500–750°C, is advantageous for improving the yield of MoS$_2$ nanooctahedra.[1,2,14] In contrast, substantial amounts of larger multi-walled quasi-spherical or nanotubular structures were produced by oven-driven reactions, but without nanooctahedra. Most likely the internal energy of the process was insufficient for the required transformation,[15–20] the nature of which has not been elucidated to date.

The confluence of these results, and the success of purely photothermal PLA, prompted the development of alternative optical systems (such as solar and discharge lamp concentrators),[21] and culminated in the development of a high-irradiance solar furnace[22] shown in Figure 2. This experimental facility provides high power densities (ca. 15 000 suns, i.e., 15 W mm^{-2}) on sizable areas (of the order of several mm^2)

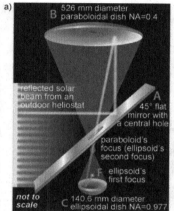

Figure 2. High-irradiance solar furnace. a) Schematic. b) Side photograph. Ambient sunlight is reflected into the laboratory from an outdoor flat dual-axis tracking mirror, and redirected by a flat 45° mirror (**A**) to a downward-facing specular paraboloidal dish (**B**) with its focus just below the tilted mirror. Flux concentration is boosted to ca. 15 000 suns by a specular ellipsoidal dish (**C**) of high numerical aperture NA at its proximate focus **F** (attainable flux density is proportional to NA^2). The quartz ampoule was inserted to **F** horizontally from the right, and translated horizontally by 1 mm each minute toward irradiating as much of the distributed precursor powder as possible.

in a continuous fashion. Relative to PLA systems, the solar furnace provides a larger reaction volume and higher vapor pressure of reactants, with peak reactor temperatures >2700 K.[23] Thermal radiation from the heated precursor powder provides a natural, extensive, and hot annealing environment.

Figure 3 shows representative nanoparticles obtained by ablating MoS$_2$ powder in a sealed, evacuated quartz ampoule in the above mentioned solar furnace. These MoS$_2$ nanostructures likely start growing from the inside out (Figure 3 a–e). This growth mode is typical for a nucleation and growth mechanism from the vapor phase,[24] and differs markedly from the outside–in growth mode where molybdenum oxide nanoparticles serve as a template for the reaction with H$_2$S.[18] When the nanooctahedra exceed a critical size, the layers should start to fold evenly and transform into quasi-spherical shape. The TEM images reveal multi-walled nanooctahedra at the core up to a diameter slightly above 10 nm or ca. 10^5 atoms beyond which the elastic strain is better accommodated by adopting an even folding of the layers into quasi-spherical shells. Tilted structures (Figure 3 c,d) reveal the inner core morphology, which is similar to that reported earlier for the smaller nanooctahedra.[1,4] The size of the hybrid nanoparticles' hollow core is determined by the maximum strain the MoS$_2$ nanooctahedra can accommodate, i.e., their minimum size, which is ca. 3 nm.[1,4]

A detailed phenomenological model that incorporates semi-empirical density functional methods[1,2] and has been shown to account for experimentally observed MoS$_2$ nanostructures ranging from fundamentally small (ca. 3 nm) to large closed-cage fullerene-like and nanotubular configurations (diameters up to 100 nm) was adopted for gaining insight into, and predicting, nanostructure stability and electronic properties. The predicted electronic properties, derived from the structural features, showed that the lattice defects associated with the formation of the nanooctahedra may induce a metallic-like nature, while quasi-spherical structures are semiconducting.

This model was used here to assess the relative stability of hollow hybrid MoS$_2$ nanoparticles comprising octahedral (subscript o) and spherical (subscript s) shells by estimating the energy relative to the monolayer ΔE. For each individual shell i of diameter D_i, where N denotes the number of atoms in a given shell, a is the lattice parameter for a MoS$_2$ monolayer (0.316 nm),[25] β is the elasticity factor (1370 eV),[2] and $\Delta\varepsilon_r$ and $\Delta\varepsilon_p$ are the energies of the atoms at the edges and corners of the nanooctahedra (1.45 and 3.72 eV per atom, respectively).[25]

$$\Delta E_s = \frac{\sqrt{3}a^2\beta}{6\pi D_s^2}; \quad N_s = \frac{6\pi D_s^2}{\sqrt{3}a^2} \qquad (1)$$

$$\Delta E_o = \frac{3\Delta\varepsilon_r(\sqrt{2}aD_o^2 - 6a^2) + 18a^2\Delta\varepsilon_p}{D_o^2}; \quad N_o = \frac{2D_o^2}{a^2} \qquad (2)$$

The energy of the full hollow hybrid nanostructure comprising o nested nanooctahedra inserted into s spherical shells is

$$\Delta E = \frac{\sum_s \Delta E_s + \sum_o \Delta E_o}{\sum_s N_s + \sum_o N_o} + \frac{k-1}{k}\varepsilon_{\mathrm{vdW}}, \quad o + s = k, \quad D_{i+1} - D_i = \frac{c}{2} \qquad (3)$$

where c is the lattice parameter for bulk 2H-MoS$_2$ (1.23 nm),[26] and $\varepsilon_{\mathrm{vdW}}$ is the van der Waals energy of interlayer interaction (-0.2 eV per atom).[25] A simplifying assumption was that the outer MoS$_2$ walls are perfectly spherical, which neglects the contribution of the cusps where the folding is uneven. This simplification is deemed accurate enough for this investigation, given the agreement between model predictions and experimental observations (see below).

The predictions plotted in Figure 4 confirm that the model can indeed account for the coexistence of nested MoS$_2$ octahedra and peripheral spherical shells, with a smooth transition from nanoclusters dominated by octahedra to those dominated by spherical shells as the particle size and number of layers increase. For a fixed total number of layers (taken to be 10, 20, and 30) and nanoparticle diameter (ranging from 13

Chapter 4
PREAMBLE FOR THE APPLICATIONS

Already in 1993 I had realized that the spherical shape of the IF-MoS_2 and WS_2 nanoparticles and their highly crystalline nature offers them a unique opportunity to serve as nanoball-bearings, i.e. as superior solid lubricants. In order to pursue this potential application, we started looking for methods to scale-up the production of the IF nanoparticles. As mentioned in **Ch. 2**, we first tried the large scale synthesis of the IF-MoS_2 NP (1.4). However, we found this objective to be illusive, due largely to the volatility of the MoO_3 at temperatures higher than 750°C, and the difficulty in controlling the complex multi-step synthesis. Therefore we turned our attention to elucidating the growth mechanism of IF-WS_2 nanoparticles from WO_3 and H_2S in a reducing atmosphere. Once this mechanism was well understood (2.1, 2.2), the next step was to go for vertical reactors. Already at this stage we were able, using the horizontal reactors, to produce 1g of the NP in one week of work. With this in hand, we approached Dr. Lev Rapoport of the Holon Institute of Technology, an expert in tribological measurements. Together with his team we were able to show the favorable effect of the IF nanoparticles as solid lubricants in different oils (4.1, 4.2). In a major recent development we were able to show that doping the IF nanoparticles with a hundred ppm of rhenium atoms and below could improve the electrical conductivity of the NP, leading to a friction coefficient in the range of 0.015 under harsh conditions, and negligible wear (2.14).

The next goal was to further scale-up the synthesis of IF-WS_2 from WO_3 powder. In the late 90's it became evident to us that only semi-continuous batch processes, like the fluidized bed reactor are suitable for this job. My students, Yishai Feldman and Alla Zak, invested a great deal of effort to get the first falling-bed reactor operating smoothly, and subsequently the first generation of the fluidized bed reactor started to produce 10–20 g/day of IF-WS_2 NP (2.5) (see **Fig. 4.1a**). The most daunting technical obstacle was feeding small amounts of the WO_3 powder (about 50 mg/min) continuously to the reactor which operated at 850°C or higher, and in a very toxic gas atmosphere (see **Fig. 4.1b**).

Once this process became routine we could perform extensive tribological experiments in collaboration with the Rapoport group and elsewhere. In a series of works with Lev's laboratory in the Holon Institute of Technology we showed

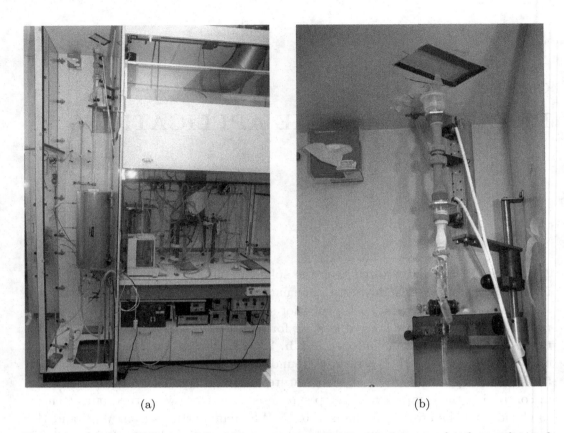

Fig. 4.1. (a) Overall image of the high temperature fluidized bed reactor for the synthesis of IF/INT of WS_2 erected in 2000. The left vertical hood hosts the reactor itself and the low rate (50 mg/min) feed system for the WO_3 powder. The hood on the right hand side hosts the H_2S cylinder; the purging system for the exit gases; the power supply and temperature control (below the right hood). In the background (behind the scale) another vertical oven is observed; (b) Blow-up of the two stage electromagnetic vibrator which feeds the reactor with WO_3 powder at a constant rate of 50 mg/min.

that we can impregnate these nanoparticles into porous metal frameworks and obtain self-lubricating surfaces (4.5). Self-lubricating polymer nanocomposites with impregnated IF-WS_2 nanoparticles were demonstrated, and this work was published in the same journal in 2004. Concomitantly, "NanoMaterials" Ltd ("ApNano Materials" Inc) was founded by Mr. Aharon Fuershtein and my former post-doc Dr. Menachem Genut. Gradually, the company started to scale-up the synthesis of the IF-WS_2 nanoparticles by constructing larger and larger fluidized bed reactors. By 2007 they were producing at a level of 80 kg/day of industrial grade IF-WS_2 nanoparticles. A new reactor with a capacity of 150 kg/day of industrial grade IF nanoparticles just began operation. Sales of the products started soon afterwards in different markets.

In collaboration with Prof. Wei-Xiang Chen of Zhejiang University in Hangzhou, China, the first electroless nickel-phosphorous coatings with impregnated IF-WS_2 NP were prepared (4.5). These hard coatings showed low friction and wear even in the absence of lubricating fluids, i.e. self-lubricating effect. In 2005 two physicians from the Faculty of Dentistry in the Hadassah Medical School of the Hebrew University in Jerusalem Dr. Meir Redlich and DMD Alon Katz knocked on my door. They told me that they heard of my work and wanted to suggest a truly new application for the IF nanoparticles. They hypothesized that if we could find a way to coat orthodontic wires with the IF nanoparticles, we could reduce the large static friction between the wires and brackets. This could possibly relieve the extraneous forces applied on the peridental (back) teeth of the patient. They hypothesized that a substantial relieve in the applied force could accelerate the teeth-healing process and thereby improve the orthodontic treatment. At the same time this coating could reduce the risk of damage to the back teeth. Alon decided to pursue his MSc thesis under the joint supervision of Meir and myself. This was the first of a number of such collaborations between my laboratory and various physicians which continues today. In his work, Alon prepared nickel coatings impregnated with the IF-WS_2 nanoparticles by electroless deposition. This work (4.7) heralded a series of papers describing metallic coatings which are both hard and self-lubricating. Such coatings may find numerous applications once fully developed.

By playing with the conditions in the fluidized bed reactor, the student Alex Margolin jointly with Dr. Rita Rosentsveig found a way to obtain substantial amounts of WS_2 nanotubes (INT) mixed with the IF nanoparticles in the fluidized bed reactor. In rare cases she could obtain phases 10% rich with the INT [1] (2.6). In 2008, Dr. Alla Zak at "NanoMaterials" started to pursue the scaling-up of the INT-WS_2 synthesis using a 1 kg reactor. Within a short time she found the conditions to synthesize a few hundred grams of an almost pure WS_2 nanotube phase (2.11). The multiwall nanotubes were quite perfect and showed very favorable mechanical behavior. No less important was the fact that by further improvements she could make the nanotubes easily dispersible (2.11) and Zak et al. [2]. This development had an immediate impact on the ability to prepare high quality reinforced polymer with a very large number of potential applications.

Now that "NanoMaterials" and my laboratory had produced sufficient amounts of high-quality IF nanoparticles/INT from WS_2, extensive work to reinforce polymer nanocomposites was initiated by different laboratories. Foremost among them is the laboratory of Dr. Mohammed Naffakh and co-workers in Madrid. Starting off in 2008, he published a long series of papers demonstrating the unique favorable contribution of the IF-WS_2 nanoparticles on the crystallization kinetics, high-temperature stability and mechanical properties of thermoplastic polymers, like isopropylene; polyphenylenesulfide and others. More recently he started studying the effect of adding carbon nanotubes or carbon fibers and IF-WS_2 nanoparticles to a variety of thermoplastic polymers, like poly (ether ether ketone) (PEEK) (2.10). A clear synergy between these two nano-additives was demonstrated. Here the lubricating effect of the IF nanoparticles led to an even better dispersion of the CNT in the polymer blend and to reducing the size of the carbon nanotube bundles. This synergy led to improved mechanical behavior of the polymer nanocomposites which could not be reached by adding any one of the NP separately. Independently and concomitantly, Professors Hanna Dodiuk and Sam Kenig and their students of careful cleaning prior to the analysis. Therefore, the carbon feature could be attributed to partial polymerization of the lubrication fluid at the elevated temperatures which develop during the wear tests. Fig. 3 shows the W(4f-5p) peaks of the surfaces within the wear tracks of IF–WS$_2$ (a) and 2H–WS$_2$ (b). The quantity of oxide is substantially higher on the surface of the wear track in contact with 2H–WS$_2$ platelets. The relative intensities I_{sul}/I_{ox}, were found to be 1.67 for the IF nanoparticles while the intensity ratio was 0.08 for the platelets. XPS spectra of the solid lubrication particles after the wear test are shown in Fig. 4. Deconvolution of the W(4f), Mo(3d) and the S(2p) spectra permitted estimation of the fraction of oxidized vs. unoxidized (native) elements in platelets and hollow nanoparticles (Fig. 4a,b). This analysis exhibits appreciably larger oxidation for the 2H than for the IF nanoparticles. The I_{sul}/I_{ox} ratio for W in platelets was 1, while negligible W oxidation was observed in the nanoparticles (0.09). For Mo this ratio was 0.16. For S the ratio was 0.03, 0.67 and 0.26 for IF–WS$_2$, 2H–WS$_2$, and 2H–MoS$_2$, respectively (Fig. 4c,d). The ratio of S/W intensities for the unoxidized elements was 2 for both IF and 2H platelets. Some spread (ca. 10%) of the data was obtained for the platelets due to an error associated with the deconvolution procedure. The XPS analysis revealed that the oxidized solid lubricant stuck to the metal surface and some of it was removed to the edge of the wear track. The fact that only a small amount of the original IF powder in the oil bath was oxidized, suggests that the oxide sticks very efficiently to the metal surface. This leads to an effective separation of the oxide from the native IF material and enhances the oil lifetime relative to the oil containing 2H–WS$_2$ or 2H–MoS$_2$ particles.

The oxidation temperatures were found to be about 320°C and 420°C for IF nanoparticles and WS$_2$ platelets (4 μm), respectively. This result indicates a greater degree of chemical stability for the macroscopic (4 μm) platelets than for the IF nanoparticles. As anticipated, platelets of diminished sizes (ca. 0.5 μm), exhibited higher chemical reactivity and decreased oxidation temperature (250°C).

It is to be noted that the experiments with large PV values were performed with samples that were provided by the scaled-up reactor (reactor II-2g/batch and above), which yielded material of poorer structural characteristics. Furthermore, detailed postmortem analysis of the powder and the wear track was limited to the low PV experiment only.

4. Discussion

4.1. Advantages of IF nanoparticles as solid lubricants

The main advantages of IF over their crystalline platelet (2H) particles are their spherical shape and the absence of dangling bonds (edge effects). This picture is obviously oversimplified, since most of the nanoparticles obtained through the present synthetic route do not have a regular spherical shapes. Furthermore, the number of atoms in each IF particle is not the same in two consecutive atomic S–M–S layers, which leads to an incommensurate layered structure and henceforth also a built-in mechanism for generation of dislocations.

The desirable tribological properties of graphite and transition-metal dichalcogenide films has been attributed to the weak van der Waals forces operating between the layers, which leads to easy shear of the films with respect to each other [15,16]. This shear mechanism is not likely to control the tribological properties of IF–WS$_2$ powder, as appreciable shear between the inner and outer closed layers is not possible in this case. Rather, their effectiveness as lubricant likely arises from several factors.

The IF particle size of typically > 100 nm is sufficient to prevent asperity contact between the mating metal surfaces. Ideally, the spherical shape of IF opens the possibility for an effective rolling friction mechanism. Further, elastic as opposed to inelastic deformations diminish the energy dissipation associated with friction and wear, thereby reducing the contact temperature. The hollow cage structure of the IF imparts a high elasticity which augments their resilience in a specific loading range [17]. Moreover, the absence of dangling bonds (edge effects) makes the IF powder chemically inert, so the nanoparticles have a lower tendency to stick either to the substrate, or to one another. It is also unlikely that the IF powders clumped and compressed into a high shear strength layer similar to that observed for C_{60} and C_{70} fullerenes [9]. Thus, it may be assumed that the mechanism of lubrication in the case of IF nanoparticles is by rolling rather than sliding friction.

Chemical effects can also play an important role in wear protection by solid lubricants. The chemical reactions that are relevant to wear of platelet materials occur predominantly at the prismatic edges, where reactive dangling bonds exist. The presence of unsaturated or dangling bonds in metal dichalcogenides leads to oxidation of the surface in the surrounding environment, especially at elevated temperatures which may occur as a result of friction. For instance, a switch from an environment of dry nitrogen to humid air led to an increase of the friction coefficient of 2H–WS$_2$ from 0.03–0.04 to 0.15–0.20 and a decrease in its resistance to oxidation by several orders of magnitude [18]. The present XPS analysis confirms the slow oxidation of IF powder and the wear track as opposed to the relatively fast oxidation when 2H platelets were used (under low PV conditions). The absence of dangling bonds may therefore be one of the prime advantages of IF nanoparticles over the crystalline platelet (2H) particles for reduction of friction and wear. The fact that IF particles oxidize at lower temperatures than the macroscopic 2H platelets is a manifestation of the strain which is involved in the deformation of the S–M–S layers during folding. It is to be noted however, that as the 2H platelets are burnished and ground during the wear tests, their diminished size leads to higher reactivity and appreciably lower oxidation temperatures as verified in the present experiments.

Under low PV conditions, the IF outperformed the 2H platelets, irrespective of the production method (reactor I or II). It is believed that under such circumstances the shape and the density of dislocations in the IF nanoparticles do not lead to their fast deformation and destruction.

4.2. The limitations of currently available IF nanoparticles

The superior tribological properties of IF nanoparticles are conceptually associated with their spherical shape and the absence of dangling bonds. It was assumed that the hollow-cage structure imparts high endurance against plastic deformation, which augments their resilience over a wide loading range. In reality, high contact pressures are realized in the ring-block scheme even at low load of 150 N. The maximal Hertzian pressure, σ_H, in the ring-block linear contact was found to be about 220–620 MPa ($\sigma_H = 0.418\sqrt{(PE)/(lr)}$, $P = 150$–1200 N). In fact, at a load of 300 N ($PV = 132$ Nm/s), many of the particles appeared to be deformed into ovoid shapes. Even so, they maintain some advantage as solid lubricants. In the subsequent experiments, increasing loads led to gradual destruction of the IF nanoparticles and formation of unsaturated bonds similar to those observed with typical platelet (2H) lubricants, which adversely affected their tribological properties. The stability of IF nanoparticles is determined not only by the exerted load but also by the contact temperature. Higher sliding velocities induce higher contact temperatures that adversely affects the tribological behavior of both IF and 2H particles, even under relatively low loads (see Table 2).

The contact surface temperature was evaluated in accordance with Archard's model [19]:

$$\Delta T_{av} = 0545 K \mu P_h^{0.75} \left[\frac{V}{k\rho c}\right]^{1/2} \left[\frac{W}{N}\right]^{1/4}$$

where the Peclet number $L > 5$ is assumed to be valid in the present case, and $K =$ partition factor, $\mu =$ coefficient of friction (Table 2), $P_h = \sigma_h$ in present case, $V =$ sliding velocity, $W =$ load, $k =$ thermal conductivity, $\rho =$ density, $c =$ specific heat, and $N =$ number of contact. In Archard's model, N is taken as unity. K has been taken to be 1, for the steel–steel pair.

Analysis of surface temperature showed that in the cases when IF nanoparticles revealed excellent tribological properties, $T_{av} < 300°C$. In the cases where the tribological properties of IF are reduced ($P = 600$ N, $V = 0.44$ m/s; $P = 300$ N, $V = 1$ m/s), T_{av} was calculated to be about 400°C. In this regard, when the temperature of the surface in contact with the fullerene-like particles is higher than their oxidation temperature (320°C), the advantage of IF nanoparticles as solid lubricants is diminished. It is expected that the synthesis of more perfectly round IF particles free of dangling bonds (dislocations) will lead to an increase of the oxidation temperature and thereby to an increase of the useful loading range for tribological applications.

The major impediment for the favorable tribological influence of IF nanoparticles under high loads and velocities is believed to emanate from the limited shape and size control that the present synthetic route offers. Smaller (ca. 30 nm) and more nearly spherical IF nanoparticles are likely to exhibit superior rolling, lower affinity to the metal surface, decreased contact temperatures, higher elasticity, higher chemical resilience, etc. Furthermore, since MoS_2 is known to exhibit improved tribological properties compared with WS_2, it is believed that IF–MoS_2 will outperform the present IF–WS_2 solid lubricant. These points will require verification once bulk synthesis of IF nanoparticles through the gas phase synthesis is advanced.

On the other hand, as the 2H particles are burnished and ground during the machinery operation, the resulting smaller particles expose more dangling bonds to the ambient. This effect leads to faster oxidation and stronger adhesion to the metal surface, which has a deleterious effect on their tribological behavior. This self-accelerating deterioration mechanism of 2H particles is in sharp contrast to the size effect in IF particles, where smaller sizes are expected to impart better tribological behavior to the mating surfaces. Preliminary data tend to strongly substantiate this hypothesis.

5. Conclusions

(1) IF–WS_2 nanoparticles appear to have excellent tribological properties within a definite loading range (PV ~ 150 Nm/s) in comparison to typical metal dichalcogenides. The oxidation of the IF particles and the wear track was less than with solid lubricants made of platelets of the same chemical compound (WS_2). The main advantages of IF nanoparticles lie in their round shape and the absence of dangling bonds.

(2) With increasing PV, deformation and destruction of IF nanoparticles diminish their tribological properties.

(3) The current bulk synthesis of IF–WS_2 through the gas–solid reaction does not yield suitable size and shape control of the nanoparticles, which adversely affect the tribological performance of IF nanoparticles, especially under high loads and sliding velocities.

(4) Application of IF nanoparticles under high loads and sliding velocity would require improved synthetic procedures (gas phase reactions), which will yield smaller and essentially spherical nanoparticles.

Acknowledgements

This research was partially supported by the UK-Israel Binational S and T Foundation and ACS-PRF (USA).

References

[1] F.P. Bowden, D. Tabor, The Friction and Lubrication of Solids, Part II, Oxford Univ. Press, London, 1964.
[2] B. Bhushan, B.K. Gupta, Handbook of Tribology, McGraw-Hill, New York, 1991.
[3] A.L. Black, R.W. Dunster, J.V. Sanders, Wear 13 (1969) 119–132.
[4] J. Gansheimerand, R. Holinsky, Wear 19 (1972) 439–449.
[5] F.P. Bowden, D. Tabor, Friction: An Introduction to Tribology, Vol. 91, Anchor, Garden City, New York, 1973.
[6] I.L. Singer, in: I.L. Singer, H.M. Pollock (Eds.), Fundamentals of Friction: Macroscopic and Microscopic Processes, Kluwer, Dordrecht, 1992.
[7] B. Bhushan, B.K. Gupta, G.W. Van Cleef, C. Capp, J.V. Coe, Appl. Phys. Lett. 62 (1993) 3253–3255.
[8] S.E. Campbell, G. Luengo, V.I. Srdanov, F. Wudi, J.N. Israelachvili, Nature 382 (1996) 520–522.
[9] P.J. Blau, C.E. Haberlin, Thin Solid Films 219 (1992) 129–134.
[10] R. Tenne, L. Margulis, M. Genut, G. Hodes, Nature 360 (1992) 444–445.
[11] Y. Feldman, E. Wasserman, D.J. Srolovitz, R. Tenne, Science 267 (1995) 222–225.
[12] L. Rapoport, Y. Bilik, M. Homyonfer, S.R. Cohen, R. Tenne, Nature 387 (1997) 791–793.
[13] Y. Feldman, G.L. Frey, M. Homyonfer, V. Lyakhovitskaya, L. Margulis, H. Cohen, G. Hodes, J.L. Hutchinson, R. Tenne, J. Am. Chem. Soc. 117 (1996) 5362–5367.
[14] E. Rabinowicz, Product Eng. 19 (1958) 31–70.
[15] L.E. Seitzman, R.N. Bolster, I.L. Singer, J.C. Wegand, Tribology Trans. 38 (1995) 445–451.
[16] J. Moser, F. Levy, Thin Solid Films 228 (1993) 257–260.
[17] D.J. Srolovitz, S.A. Safran, M. Homyonfer, R. Tenne, Phys. Rev. Lett. 74 (1995) 1779–1881.
[18] S.V. Prasad, J.S. Zabinski, J. Mater. Sci. Lett. 11 (1993) 1413–1415.
[19] J.F. Archard, Wear 2 (1958) 438–455.

Wear and Friction of Ni-P Electroless Composite Coating Including Inorganic Fullerene-WS$_2$ Nanoparticles**

By Wei Xiang Chen*, Jiang Ping Tu, Zhu De Xu, Reshef Tenne, Rita Rosenstveig, Wen Lu Chen, and Hai Yang Gan

It was proposed that nanoparticles of layered compounds are not stable in the planar form and they spontaneously fold into fullerene-like structures and nanotubes (IF phases),[1,2] which are analogous to the carbon fullerenes[3] and carbon nanotubes.[4] This concept has been transcended now to materials with low dimensionality. Thus, nanotubes and fullerene-like particles from various layered compounds have been reported in the literature in recent years.[5] Nanotubes and fullerene-like nanoparticles of the superconducting phases NbS$_2$ and TaS$_2$ have been recently synthesized.[6–8] MoS$_2$ nanooctahedra, which can be considered the analogs of C$_{60}$ in this compound have been reported.[9] Also recently, single wall MoS$_2$ nanotubes with a diameter of less than 1 nm have been reported.[10] Substantial amounts of the pure fullerene-like MoS$_2$ phase and MoS$_2$ nanotubes have been prepared as early as 1995.[11] It is well known that powders of the 2H polytype of MoS$_2$ (2H-MoS$_2$) and WS$_2$ (2H-WS$_2$) with micron size particles have been widely applied as solid lubricants.[12] However, these particles tend to stick to the underlying metal surfaces through their prismatic (hk0) faces, which leads to deleterious mechanical and chemical degradation of the solid lubricant. Considering the spherical topology of the IF nanoparticles, it was proposed that they could ideally suit tribological applications. In this paper, IF-WS$_2$ nanomaterials were synthesized employing WO$_3$ powders as the precursor by a solid-gas reaction, and characterized by X-ray diffraction, scanning electron microscopy and transmission electron microscopy. Ni-P-(IF-WS$_2$) electroless composite coating was prepared from a suspension of IF-WS$_2$ nanoparticles in electroless bath. The results demonstrated that the Ni-P-(IF-WS$_2$) exhibited better tribological performances than Ni-P-(2H-WS$_2$) and Ni-P-graphite electroless composite coatings. The mechanism of improvement of the tribological properties of the electroless composite coating was also discussed.

It was reported that mixing a few percents of IF-WS$_2$ nanoparticles in lubricating fluids led to substantial improvements in the tribological behavior of these suspensions,[13,14] especially under heavy loads and small velocities, where the boundary or mixed lubrication regimes are valid. More recently, thin film coatings of IF-MoS$_2$ nanoparticles were prepared by the arc-discharge method, and their tribological properties were tested.[15] These films exhibited excellent tribological behavior, with friction coefficients in the range of 0.005 even in humid atmosphere where sputtered MoS$_2$ films deteriorate very rapidly. An alternative strategy is presented by the work of Rapoport et al.,[16–18] who studied the tribological properties of nanocomposite structures, in which the IF nanoparticles were impregnated into porous metal matrices and slowly released with time to the metal surface. The IF nanoparticles, which were released to the metal surface provided very easy shear and also served as spacers, eliminating contact between the asperities of the mating metal surfaces, and thereby slowing the wear rate considerably. Since the nanoparticles could be impregnated into very small pores, the mechanical toughness of the metal matrix was not compromised. It was nonetheless observed that the tribological properties of the IF nanoparticles was improved by adding minute quantities of lubricating fluid, which had only a small tribological effect if added alone (without the IF nanoparti-

---

[*] Prof. W. X. Chen, Prof. Z. D. Xu, Dr. W. L. Chen, H. Y. Gan
Department of Chemistry, Zhejiang University
Hangzhou 310027 (P. R. China)
E-mail: weixiangchen@css.zju.edu.cn

Prof. J. P. Tu
Department of Materials Science and Engineering
Zhejiang University
Hangzhou 310027 (P. R. China)
E-mail: tujp@cmsce.zju.edu.cn

Prof. R. Tenne, Dr. R. Rosenstveig
Department of Materials and Interfaces
the Weizmann Institute of Science
Rehovot 76100 (Israel)
E-mail: reshef.tenne@weizmann.ac.il

[**] This work was supported by the National Natural Science Foundation of China (50171063, 20003009 and 20171039), the Israeli Ministry of Science (Tashtiot), the Israel Science Foundation and the Minerva Foundation (Munich) as well as the Zhejiang Provincial Natural Science Foundation of China (501074).

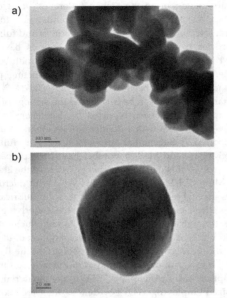

Fig. 1. A transmission electron microscopy (TEM) image of a sample of IF-WS$_2$ powder.

Fig. 2. SEM images of IF-WS$_2$ and 2H structure (platelets form) WS$_2$ powders.

cles). The favorable tribological effect of the IF nanoparticles was attributed to their ability to easily roll/slide on the metal surface. In fact, this property explains also the great difficulty to image the nanoparticles with scanning probe techniques, at least in a contact mode, where substantial shear forces are exerted by the tip. An alternative explanation was provided by Golan et al., who used the surface force apparatus for measurements of the shear forces of WS$_2$ nanoparticles in tetradecane.[19–20] They observed, that while WS$_2$ nanoparticles with 2H structure (platelets form) lead to a substantial increase in the friction coefficient of the pure tetradecane, the IF nanoparticles produced a very favorable effect.

Electroless deposition technique has been widely used to prepare metal-based composite coatings for tribological applications, such as Ni-P-SiC, Ni-P-BN and Ni-P-graphite. In general, these electroless composite coatings are expected to have higher wear resistance than the bear metal alloy coating. Since the inorganic fullerene-like nanoparticles exhibit excellent tribological properties, it is likely the electroless composite coating containing IF nanosize particles of this kind will show better tribological parameters than that of the traditional electroless composite coatings, such as Ni-P-SiC, Ni-P-graphite and Ni-P-(2H-WS$_2$). In this paper, the preparation and tribological properties of Ni-P-(IF-WS$_2$) electroless composite coating is described. For the present work, IF-WS$_2$ nanosize particles were synthesized by solid-gas reaction using WO$_3$ nanopowder as a precursor under N$_2$/H$_2$/H$_2$S gas flow. The experimental details of the synthesis were described elsewhere.[21] A transmission electron microscopy (TEM) image of an IF-WS$_2$ powder is shown in Figure 1.

Obviously, the shapes of this IF-WS$_2$ nanoparticles is very close to a sphere, and has nested fullerene-like structure, which is analogous to that of carbon nano-onions. Figure 2 shows scanning electron microscopy (SEM) images of IF-WS$_2$ and 2H-WS$_2$ powders. It can be seen from Figure 2 that the entire IF-WS$_2$ powder consists of nanoparticles with morphologies close to spheres, and diameter in the range of 100–200 nm. On the other hand, the 2H-WS$_2$ powder consists of platelet ranging in sizes between 600–1700 nm. X-ray diffraction (XRD) patterns of IF-WS$_2$ and 2H-WS$_2$ powders are shown in Figure 3. Most of the XRD peaks of the IF-WS$_2$ powder coincide with these of 2H-WS$_2$. However, in comparison with the XRD patterns of IF-WS$_2$ and 2H-WS$_2$ some differences are discernable, like the shape and relative intensity of the peaks. Most importantly, the (002) peak of the IF nanopowder is shifted to lower angles, compared to the (002) peak of the 2H polytype. This shift is caused by a lattice expansion in the IF powder of about 2–3 % along the c-axis compared to the bulk (2H) form.[11] As shown in Figure 3, the FWHM of the diffraction peaks of IF-WS$_2$ are larger than that of 2H-WS$_2$. For the (002) peak, the FWHM of IF-WS$_2$ is 0.545 degree, while that of 2H-WS$_2$ is 0.158 degree, suggesting size difference of about one order of magnitude between the particles of the two phases.

Ni-P-(IF-WS$_2$) composite coating was deposited on the surface of a mild carbon steel substrate prepared from a suspension of IF-WS$_2$ nanoparticles in aqueous bath containing nickel sulphate and sodium hypophosphite by electroless plating method. Figure 4 shows XRD patterns of as-prepared Ni-P-(IF-WS$_2$) and Ni-P-(2H-WS$_2$) composite coatings. These

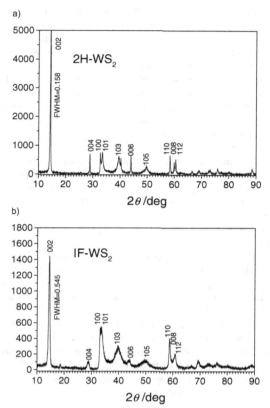

Fig. 3. X-ray diffraction patterns of IF-WS$_2$ and 2H-WS$_2$ powders.

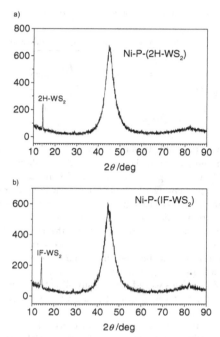

Fig. 4. XRD patterns of as-prepared Ni-P-(IF-WS$_2$) and Ni-P-(2H-WS$_2$) electroless composite coatings.

patterns indicate that the microstructure of the as-prepared electroless composite coatings are mainly amorphous. As shown also in Figure 4, the (002) peak of IF-WS$_2$ and 2H-WS$_2$ can be obviously found in Ni-P-(IF-WS$_2$) and Ni-P-(2H-WS$_2$) composite coating, respectively.

In order to improve the mechanical and tribological properties of electroless composite coatings, they were annealed for 2 h at 673 K in a vacuum furnace. As shown in Figure 5, after annealing at 673 K, the microstructure of the electroless composite coating changed from amorphous to crystalline structure, and Ni$_3$P phase was precipitated. The wear rate and friction coefficient for the electroless composite coatings were measured using a ring-on-block test rig under oil lubrication. The results of the tribological testing are summarized in Table 1. It is found that the Ni-P-(IF-WS$_2$) composite coating exhibits both the highest wear resistance and the lowest friction coefficient among these electroless coatings. It is accepted that the tribological behavior of the electroless coatings can be improved by dispersing some inorganic particles in the coating. As shown in Table 1, the Ni-P-(2H-WS$_2$) and Ni-P-graphite composite coatings present higher wear resistance and lower coefficient than the reference Ni-P coating. It may be noted that the favorable effects of IF-WS$_2$ nanoparticles on the wear rate and friction coefficient are larger than those of 2H-WS$_2$ and graphite for electroless composite coatings. Figure 6 shows SEM images of the worn surface of electroless coatings. The surface of the tested Ni-P-(IF-WS$_2$) coating appears to be rather smooth, in which only slight worn scars can be observed on the worn surface of the composite coating, while apparent wear scars can be found on the Ni-P, Ni-P-(2H-WS$_2$) and Ni-P-graphite surfaces. Thus, it is rather clear that IF-WS$_2$ nanoparticles can effectively prevent spalling of the composite coating.

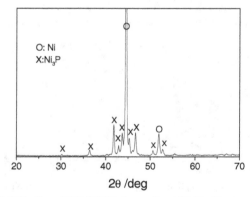

Fig. 5. XRD patterns of Ni-P-(IF-WS$_2$) after annealing at 673 K for 2 h.

Figure 2. SEM micrograph of the composite coating on the orthodontic archwire (right). The *IF* nanoparticles can be seen within the Ni–P matrix in the magnified image (left).

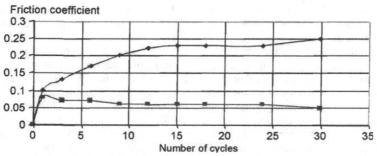

Figure 3. Tribological tests were performed using a ball-on-flat device with a sliding velocity of 0.2 mm/s and a load of 50 g (ca. 1.5 GPa). A ball bearing with a diameter of 2 mm was used as a counter body. Dry and wet friction-tests with paraffin oil lubricant were carried out during 50–200 cycles. The hardness of the Ni–P coating was close to 7500 MPa. This relatively high hardness usually led to formation of ploughing tracks on the surface of the ball. Blue curve is the uncoated stainless steel wire; red curve is the wire coated with Ni–P and *IF*-WS$_2$ nanoparticles.

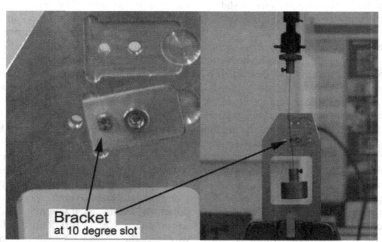

Figure 4. A photograph showing the device used to mount the brackets and orthodontic wires onto the Instron setup for the simulation of the archwire functioning in the mouth. About 12-cm segments of the orthodontic wires (coated and uncoated) were attached, on their upper part, to a 10 Newton load cell and the lower end was connected to a 150 g weight. The wires were then inserted into the slots in the brackets and ligated with an elastomeric module (Sani-Ties Silver, GAC) to all four wings of the brackets. The 150 gram weight was used to re-straighten the wire following its insertion into the bracket similar to the clinical situation.

Deionized water was used to simulate the conditions in the mouth. Table 1 summarizes the results of the mechanical measurements. Each data point is an average of five different measurements. The results show a substantial reduction in the friction resistance to sliding at the different tilt angles, both in dry and wet modes. At an angle of 0° the reduction of friction was only 17%. As the angle grew to 5°, the reduction rate grew to 46% and the 10° angle showed a 54% reduction of friction compared to the non-coated wire.

Table 1.
Summary of the results of the mechanical experiment (in N ± SD).

Angle/Coating	0°	5°	10°
Non-coated wire	1.32 ± 0.12	2.95 ± 0.09	4.00 ± 0.19 dry
			3.35 ± 0.21 wet
Ni–P + *IF* coated wire	1.10 ± 0.06	1.58 ± 0.25	1.85 ± 0.21 dry
			1.57 ± 0.23 wet

The mechanism by which this reduction is achieved can be explained by the theories suggested by Rapoport et al. [15]. At the first stage when there is no angle between the slot and wire, the *IF* nanoparticles act as spacers and reduce the number of asperities that come in contact, resulting in a lower coefficient of friction. As the angle grows the load at the edges of the slot increases causing the higher friction at the uncoated wire. It is probably at this point on the coated wire that the release of *IF* nanoparticles from the coating into the tribological interface and their exfoliation occurs, resulting in the formation of a solid lubricant film on the sliding wire. The higher load at this point brings the asperities of the mating surfaces in straight contact causing the fluid (saliva in the mouth) to be squeezed out of the gap between the wire and slot, relying on the excellent tribological behavior of the solid lubricant film to allow the sliding of the archwire. When the two materials are SS, as is the case with the uncoated wire, the friction coefficient is high. The presence of WS_2 nano-sheets at the interface under high loads, leads to a very facile sliding between these sheets thereby reducing the coefficient of friction.

Due to the tipping and uprighting type of tooth movement that is encountered during orthodontic treatment, this type of lubrication is most desirable because the main problem of resistance to sliding is found at the angles higher than the critical contact angle. A possible future significant reduction in the levels of orthodontic forces (which had never occurred during the last century) will lead this dental specialty towards a new era of faster and safer treatment. If low forces are to be used then the problem of "loss of anchorage" will no longer remain an issue.

The tribological mechanisms responsible for the improved lubricity of surfaces coated with *IF*-WS_2 nanoparticles have been investigated in quite a detail [16–18]. Surface force apparatus measurements combined with Auger and transmission electron microscopy (TEM) analysis provided evidence for the third body effect, i.e. transfer of exfoliated WS_2 nano-sheets onto the underlying surfaces [16]. Tribological tests followed by extensive TEM, SEM and X-ray photoelectron spectroscopy (XPS) analyses [17] revealed that under low pressure the *IF*-WS_2 nanoparticles remained intact at the metal-metal interface. Contrarily, under high pressures (>1 GPa) third body transfer of exfoliated WS_2 nano-sheets occurred, producing a protective film on the matting surfaces. Since the pressure in the contact area is estimated to be on the order of a few MPa the effect of third body is not likely to be very significant in the present system. Similar observations were made by analyzing surfaces lubricated with *IF*-MoS_2 [18].

Further investigation is still needed before this coating can be utilized in orthodontics. First is the biocompatibility of the coating. Preliminary toxicology tests of the *IF*-WS_2 reported the material as being non-toxic in oral administration of rats (Tsabari, H. Inorganic fullerene-like WS_2 nanospheres (*IF*-WS_2) *(Batch No.: HP6)* Acute oral toxicity, acute toxic class method in the rat: Final report. Harlan Biotech Israel Feb. 28, 2005). Additional testing by a certified laboratory has proven that the *IF*-WS_2 powder does not lead to any dermal related toxicity upon exposure to skin (skin sensitization test). More recently, inhalation tests were carried out on rats by another certified laboratory with no toxicity effects observed whatsoever. Obviously, future clinical use of the coated wires will be subjected to the safe biocompatibility tests according to accepted procedures. The presence of Ni in orthodontic appliances is known and they are approved for use (except in people with an allergic sensitivity to Ni) but the composite coating needs to be further tested. Furthermore, new self-lubricating coatings in which metals other than Ni–P were used have been recently prepared and are currently being analyzed. Since the local pressure in the contact point between the wire and bracket is not particularly high polymer coatings, which were recently prepared, could be also used for this purpose. Further studies and optimization of the self-lubricating wire coatings are now under way. Therefore, the present data can be considered as a preliminary report only. Extensive wear tests are also underway for such devices. Using saliva, or simulating fluids may lead to extensive corrosion due to electrical potential difference between the wire and the bracket. Application of the self-lubricity behavior of such coatings in other medical devices, like needles, catheters, endoscopes, and coating surfaces of articulating joints are foreseen.

1. Experimental

Orthodontic wires (Ormco, California 0.019× 0.025 inch^2 rectangular SS) were coated with a uniform and smooth Ni-P film using 100 ml solution

(ENPLATE Ni-425, Enthone Inc.). The orthodontic wires (or SS plate substrate) were inserted into the electroless Ni-P bath (88 °C, pH = 4.8, magnetic stirring) for 30 min. The plating resulted in a shiny smooth Ni–P layer. To another electroless solution 200 mg of an agglomerated *IF* powder with average particle size of 120 nm were added together with a cationic surfactant (cetyltrimethylammoniumbromide—CTAB). A short (1 min) sonication (Sonifaier 150, Branson-30 Watts) was used to disperse the agglomerates and ensure the stability of the suspension. A special procedure was developed to deposit a uniform and relatively smooth composite Ni-P + *IF* film and secure its adequate adherence to the underlying archwire. First the wires were etched with HF (20%) solution. Subsequently, Ni coating was applied using e-beam evaporation. On top of this film, a pure Ni-P electroless film was deposited, and only at the final stage of the process electroless Ni–P + *IF* film was deposited onto the etched wires (figure 2). The quality of the film was evaluated using various analytical techniques, including scanning and transmission electron microscopy, X-ray diffraction, Raman microscopy. The film did not peel-off in either a scotch-tape test or even under severe bending of the wire with a 2 cm radius of curvature.

Acknowledgment

R. Tenne is the holder of the Drake Family Chair in Nanotechnology and the director of the Helen and Martin Kimmel Centre for Nanoscale Science. H.D. Wagner holds the Livio Norzi Chair in Materials Science.

References

[1] G.A. Thorstenson and R.P. Kusy, Am. J. Orthod. Dentofacial. Orthop. 120 (2001) 361.
[2] A. Cash, R. Curtis, D. Garrigia-Majo and F. McDonald, Eur. J. Orthod. 26 (2004) 105.
[3] Z. Fuss, I. Tsesis and S. Lin, Dent. Traumatol. 19 (2003) 175.
[4] D. Roberts-Harry and J. Sandy, Br. Dent. J. 196 (2004) 255.
[5] J. Chung-Chen, L. Jang-Jaer, C. Hsing-Yu, J.C. Zwei-Chieng, C. Hsin-Fu and J. Yi., Angle Orthod. 75 (2004) 626.
[6] R.P. Kusy, Angle Orthod. 70 (2000) 366.
[7] R.P. Kusy and J.Q. Whitley, Semin. Orthod. 3 (1997) 166.
[8] S.P. Jones, C.C. Tan and H.H. Davies, Eur. J. Orthod. 24 (2002) 183.
[9] R. Tenne, L. Margulis, M. Genut and G. Hodes, Nature 360 (1992) 444.
[10] L. Rapoport, Y. Bilik, Y. Feldman, M. Homyonfer, S.R. Cohen and R. Tenne, Nature 387 (1997) 791.
[11] W.X. Chen, Z.D. Xu, R. Tenne, R. Rosensteig, W.L. Chen, H.Y. Gan and J.P. Tu., Adv. Eng. Mater. 4 (2002) 686.
[12] L. Rapoport, N. Fleischer and R. Tenne, J. Mater. Chem 15 (2005) 1782.
[13] G.O. Mallory and J.B. Hajdu, *Electroless Plating – Fundamentals and Applications* (William Andrew Publishing/Noyes, New York, 1990) 269–287.
[14] M. Redlich, Y. Mayer, D. Harari and I. Lewinstein, Am. J. Orthod. Dentofacial. Orthop. 124 (2003) 69.
[15] L. Rapoport, V. Leshchinsky, I. Lapsker, Y. Volovik, O. NepomnyashchyM. Lvovsky, R. Popovitz-Biro, Y. Feldman and R. Tenne, Wear 255 (2003) 785.
[16] C. Drummond, N. Alcantar, J. Israelachvili, R. Tenne and Y. Golan, Adv. Funct. Mater. 11 (2001) 348.
[17] L. Rapoport, V. Leshchinsky, I. Lapsker, Yu. Volovik, O. Nepomnyashchy, M. LvovskyR. Popovitz-Biro, Y. Feldman and R. Tenne, Wear 255 (2003) 785.
[18] L. Cizaire, B. Vacher, T. Le Mogne, J.M. Martin, L. Rapoport, A. Margolin and R. Tenne, Surf. Coat. Tech. 160 (2002) 282.

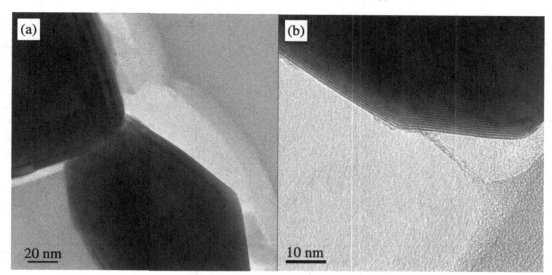

Figure 10. TEM micrograph of microtomed IF-WS$_2$ particles/epoxy resin. Mixing conditions were 18 000 rpm for 4 h at a temperature of 50°C, (a) detachment of the fullerene from the matrix, (b) strong adhesion between the particle and the matrix.

addition a wide and medium absorption peak was obtained around 1543 ν_{max}/cm^{-1} indicative of derivatives of the CO–NH–R bond. The other absorptions around 1650 and 3317 ν_{max}/cm^{-1} are indicative of OH and NH moieties (Fig. 11(b)).

From these results it appears that in the curing process a nucleophilic attack took place between the outermost sulfur atoms of the nanoparticles and the oxa-cyclopropane group of the epoxy to form a covalent bond between carbon and sulfur that can be oxidized to sulfonic or sulfuroxide. This C–S species may reduce the ability of the epoxy to react with the amide group from the curing agent and, thus, hinder the forming of cross-linked bonds.

4. Conclusions

Epoxy nanocomposites based on IF-WS$_2$ were prepared using high shear mixing in order to disperse IF-WS$_2$ nanoparticles in the epoxy resin. Shear and peel tests of the IF-epoxy nanocomposite were carried out in bonded joints followed by transmission and scanning electron microscopy of the failed surfaces. It was found that long mixing times and high sheering levels reduced the agglomeration of the IF-WS$_2$ and resulted in a homogeneous dispersion of the nanoparticles.

It was also found that only at low concentrations of the IF-WS$_2$ (0.5 wt%) did the composite material exhibited high shear and peel strength. At concentrations above this threshold, the peel strength decreased sharply. The SEM analysis of the fractured surfaces indicated that a variety of energy absorbing mechanisms took place leading to these somewhat unexpected results. FT-IR analysis verified that the sulphur in the outermost layer of the nanoparticles reacted with the epoxy group to form a C–S bond.

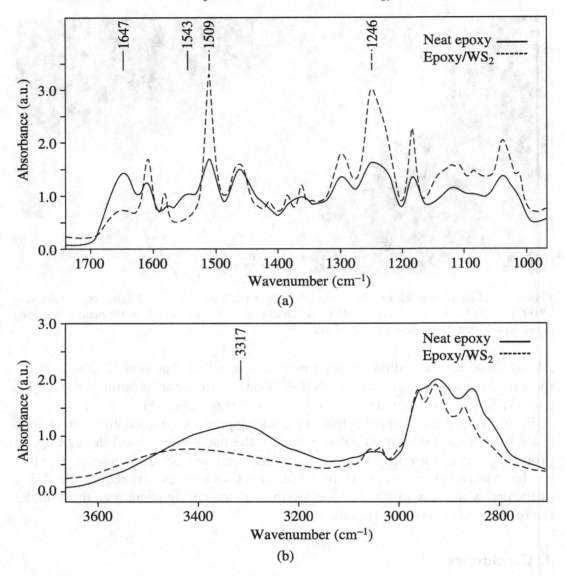

Figure 11. (a) Region of 1000–1700 cm^{-1}; (b) 2800–3600 cm^{-1} in the FT-IR spectra of the untreated and mixed epoxy films. Mixing conditions were 18 000 rpm for 4 h at a temperature of 50°C.

The simultaneous increase of both shear and peel strength at very low concentrations of the IF-WS$_2$ nanoparticles, may find applications in high performance adhesives and coatings as well as in structural and ballistic fiber composites.

Acknowledgements

We are grateful to the staff of the Plastics and Rubber Center at Shenkar College of Engineering & Design for their assistance and to NanoMaterials, Ltd. for providing the IF-WS$_2$ nanoparticles. We also thank Dr. R. Popovitz-Biro (Weimann Institute) for the assistance in the TEM analysis. We acknowledge the Irving and Cherna

Moskowitz Center for Nano and Bio-Nano Imaging at the Weizmann Institute of Science and the H. Perlman, Horowitz and the Gurwin Foundations. Reshef Tenne is the Drake Family Chair Professor of Nanotechnology and director of the Helen and Martin Kimmel Center for Nanoscale Science.

References

1. M. E. Wrigione, L. Mascia and D. Acierno, *Eur. Polvm. J.* **31**, 1021 (1995).
2. R. Bagheri and A. R. Pearson, *Polymer* **41**, 269 (2000).
3. R. J. Day, P. A. Lovell and A. A. Wazzan, *Comp. Sci. Technol.* **61**, 41 (2001).
4. H. J. Sue, E. I. Garcia Meitin, D. M. Pickelman and C. J. Bott, *Colloid Polym. Sci.* **274**, 342 (1996).
5. J. Lee and A. F. Yee, *Polymer* **42**, 589 (2001).
6. J. Lee and A. F. Yee, *J. Mater. Sci.* **36**, 7 (2001).
7. Y. Huang and A. J. Kinloch, *Polymer* **33**, 1330 (1992).
8. D. Ratna, O. Becker, R. Krishnamurthy, G. P. Simon and R. J. Varley, *Polymer* **44**, 7449 (2003).
9. R. J. Varley and W. Tian, *Polym. International* **53**, 69 (2004).
10. B. J. Cardwell and A. F. Yee, *J. Mater. Sci.* **33**, 5473 (1998).
11. R. Bagheri and R. A. Pearson, *Polymer* **37**, 4529 (1996).
12. K. Kozio, J. Vilatela, A. Moisala, M. Motta, P. Cunniff, M. Sennett and A. H. Windle, *Science* **318**, 1892 (2007).
13. D. Eder and A. H. Windle, *Adv. Mater.* **9**, 1787 (2008).
14. M. Zhang, S. Fang, A. A. Zakhidov, S. B. Lee, A. E. Aliev, C. D. Williams, K. R. Atkinson and R. H. Baughman, *Science* **309**, 1215 (2005).
15. B. Wetzel, P. Rosso, F. Haupert and K. Friedrich, *Eng. Fracture Mechanics* **73**, 2375 (2006).
16. A. S. Argon and R. E. Cohen, *Polymer* **44**, 6013 (2003).
17. R. Tenne, L. Margulis, M. Genut and G. Hodes, *Nature* **360**, 444 (1992).
18. Y. Feldman, E. Wasserman, D. J. Srolovitz and R. Tenne, *Science* **267**, 222 (1995).
19. Y. Feldman, G. L. Frey, M. Homyonfer, V. Lyakhovitskaya, L. Margulis, H. Cohen, G. Hodes, J. L. Hutchison and R. Tenne, *J. Am. Chem. Soc.* **118**, 5362 (1996).
20. R. Tenne, *Nature Nanotechnology* **1**, 103 (2006).
21. A. Zak, L. Sallacan-Ecker, A. Margolin, M. Genut and R. Tenne, *Nano* **4**, 91 (2009).
22. L. Rapoport, Y. U. Bilik, Y. Feldman, M. Homyonfer, S. R. Cohen and R. Tenne, *Nature* **387**, 791 (1997).
23. L. Rapoport, O. Nepomnyashchy, R. Popovitz-Biro, Y. Volovik, B. Ittah and R. Tenne, *Adv. Eng. Mater.* **6**, 44 (2004).
24. M. Naffakh, Z. Martin, N. Fanegas, C. Marco, M. A. Gomez and I. Jimenez, *J. Polym. Sci. Part B: Polym. Phys.* **45**, 2309 (2007).
25. M. Naffakh, C. Marco, M. A. Gomez and I. Jimenez, *J. Phys. Chem. B* **112**, 14819 (2008).
26. X. Hou, C. X. Shan and K. L. Choy, *Surface & Coatings Technology* **202**, 2287 (2008).
27. I. Kaplan-Ashiri, S. R. Cohen, K. Gartsman, V. Ivanovskaya, T. Heine, G. Seifert, I. Wiesel, H. D. Wagner and R. Tenne, *Proc. Natl. Acad. Sci.* **103**, 523 (2006).
28. Y. Feldman, A. Zak, R. Popovitz-Biro and R. Tenne, *Solid State Sci.* **2**, 663 (2000).
29. S. Komarneni, *J. Mater. Chem.* **2**, 1219 (1992).
30. Z. Bartczak, A. S. Argon, R. E. Cohen and M. Weinberg, *Polymer* **40**, 2347 (1999).
31. B. Wetzel, F. Hauperta and M. Q. Zhangb, *Comp. Sci. Technol.* **66**, 2055 (2003).
32. S. Y. Fu, X. Q. Feng, B. Lauke and Y. W. Mai, *Comp. Part B: Eng.* **39**, 933 (2008).

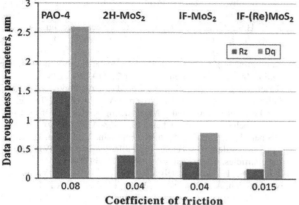

Fig. 8 A variation of the roughness parameters under friction in steady state with PAO-4 oil and different additives. The sampling length is 0.5 mm

friction with different additives at a steady friction conditions.

It may be seen that the low friction observed with Re:IF-MoS$_2$ NP is accompanied by formation of smoothest tribofilm produced on the surface of the steel disk. It is suggested that the filling of the deep valleys on the disk surface by Re:IF-MoS$_2$ NP increases the real contact area and thus decrease the effective contact pressure.

It is interesting to note that when decreasing the sampling length down to 25 μm (AFM test), the parameter Rz of the steel surfaces covered with Re:IF-MoS$_2$ NP went down to approximately 40 nm. Such smooth surfaces can be obtained because of the facile penetration of the well-distributed Re:IF-MoS$_2$ NP into the gap between rubbed surfaces. The formation of this highly smooth surface and a thick tribofilm with the Re-doped NP can be also the result of the NP surface (negative) charge of the NP. It is expected that the charged NP exert mutual repulsion at close proximity and prevent their sedimentation and agglomeration into large bulges. Furthermore, the agglomeration of the undoped IF NP, which hinder their facile access to the interface [13] is largely alleviated in the case of the doped NP. Application of n-doped NP can lead to their mutual repulsion and thus to decreasing the shear forces of the film coated surfaces. The ultra-low friction coefficient (0.015) can be also associated with rolling/sliding friction of the NP on small sampling length. Here, individual NP or small aggregates can roll/shear on the smooth surface of Re-doped film. This rolling/shearing mechanism is confirmed also in recent direct experiments with individual IF NP [16, 18]. In particular a detailed analysis of the behavior of individual IF-MS$_2$ (M = Mo, W) was undertaken under high pressure (0.5–3.5 GPa) in relation to their structure [16]. While the IF-MoS$_2$ NP are squashed, but of high crystalline order, the IF-WS$_2$ NP possess a closer to spherical shape, but exhibit more structural defects. Therefore, the IF-WS$_2$ NP show clear rolling under combined normal load/shear, but they tend to collapse earlier (app. 1–2 GPa) under the combined load/shear forces than their IF-MoS$_2$ counterparts which collapse under a load of 2–3.5 GPa. Also, the squashed IF-MoS$_2$ NP exhibit facile shearing which provides a very effective lubrication mechanism. Analysis of this kind has not been performed to individual Re-doped IF NP as yet. While rolling/shearing is likely to be more effective in the Re-doped NP, the main differences between those of WS$_2$ and MoS$_2$ are likely to remain.

4 Conclusions

(1) Doping of fullerene-like MoS$_2$ and also WS$_2$ nanoparitcles were done with rhenium atoms. The doped NP were characterized by different methods. In particular, the Re atoms at concentrations <0.1 at.% were found to substitute the molybdenum lattice atoms and donate electron to the conduction band making the NP negatively charged.

(2) The electrical resistivity versus temperature of Re:IF-MoS$_2$ pellets and bulk (2H-MoS$_2$) powder was studied. It has been shown that the specific resistivity goes down with the Re doping level. The STS shows that the density of states near the Fermi level increases with the Re doping, supporting its substitutional doping effect and the increased conductivity of the doped MoS$_2$ NP.

(3) A homogeneous dispersion of the doped NP in oil is associated with surface charges leading to mutual repulsion at close proximity which prevents sedimentation and agglomeration of the NP.

(4) Application of Re-IF-MoS$_2$ NP provides ultra-low friction coefficient and very low wear rate. The unique tribological properties of Re-doped NP can be related to rolling effect of the NP and tribo-charging of the tribofilms by the surface layer of Re:IF-MoS$_2$ NP.

5 Highlights

New synthesis of WS$_2$ and MoS$_2$ nanoparticles (NP) with fullerene-like structure doped with small amounts (< 1 at.%) of rhenium atoms has been developed. The resistivity of the NP was shown to decrease significantly with increasing doping level. The doped NP were shown to exhibit reduced agglomeration and produce stable suspensions in PAO-4 and PAO-6 oils. Application of Re-IF-MoS$_2$ NP provides ultra-low friction coefficient and very low wear rate.

Acknowledgment We are grateful to Dr. R. Popovitz-Biro for the help with the TEM analysis. RT gratefully acknowledges (a) the support of ERC project *INTIF* 226639, (b) the Israel Science Foundation, (c) the Harold Perlman Foundation, and (d) the Irving and Cherna Moskowitz Center for Nano and Bio-Nano Imaging. RT is the Drake Family Chair in Nanotechnology and director of the Helen and Martin Kimmel Center for Nanoscale Science.

References

1. Tenne, R., Margulis, L., Genut, M., Hodes, G.: Polyhedral and cylindrical structures of WS$_2$. Nature **360**, 444–445 (1992)
2. Rapoport, L., Bilik, Yu., Feldman, Y., Homyonfer, M., Cohen, S.R., Tenne, R.: Hollow nanoparticles of WS2 as potential solid-state lubricants. Nature **387**, 791–793 (1997)
3. Tannous, J., Dassenoy, F., Bruhacs, A., Tremel, W.: Synthesis and tribological performance of novel Mo$_x$W$_{1-x}$S$_2$ ($0 < x < 1$) inorganic fullerenes. Tribol. Lett. **37**, 83–92 (2010)
4. Katz, A., Redlich, M., Rapoport, L., Wagner, H.D., Tenne, R.: Self-lubricating coatings containing fullerene-like WS$_2$ nanoparticles for orthodontic wires and other possible medical applications. Tribol. Lett. **21**, 135–139 (2006)
5. Hou, X., Shan, C.X., Choy, K.L.: Microstructures and tribological properties of PEEK-based nanocomposite coatings incorporating inorganic fullerene-like nanoparticles. Surf. Coat. Techn. **202**, 2287–2291 (2008)
6. Naffakh, M., Martın, Z., Fanegas, N., Marco, C., Gomez, M.A., Jimenez, I.: Influence of inorganic fullerene-like WS$_2$ nanoparticles on the thermal behavior of isotactic polypropylene. Polym. Sci. Part B: Polym. Phys. **45**, 2309 (2007)
7. Dowson, D.: An extensive historical review of work in the area of friction and wear, dating back to prehistorical times. In: History of tribology, pp. 28–33. Professional Engineering Publishing, London (1998)
8. Park, H.S., Hwang, J., Choa, S.H.: Tribocharge build-up and decay at a slider-disk interface. Microsystem Technol. **10**, 109–114 (2004)
9. Derjaguin, B.V., Smilga, V.P.: Electrostatic component of the rolling friction force moment. Wear **7**, 270–281 (1964)
10. Rosentsveig, R., Margolin, A., Gorodnev, A., Popovitz-Biro, R., Feldman, Y., Rapoport, L., Novema, Y., Naveh, G., Tenne, R.: Synthesis of fullerene-like MoS$_2$ nanoparticles and their tribological behavior. Mater. Chem. **19**, 4368–4374 (2009)
11. Deepak, F.L., Popovitz-Biro, R., Feldman, Y., Cohen, H., Enyashin, A., Seifert, G., Tenne, R.: Fullerene-like Mo(W)$_{1-x}$Re$_x$S$_2$ nanoparticles. Chem. Asian J **3**, 1568–1574 (2008)
12. Yadgarov, L., Rosentsveig, R., Leitus, G., Albu-Yaron, A., Moshkovich, A., Perfilyev, V., Vasic, R., Frenkel, A.I., Enyashin, A.N., Seifert, G., Rapoport, L., Tenne, R.: Controlled doping of MS$_2$ (M=W, Mo) nanotubes and fullerene-like nanoparticles, Angew. Chem. Intl. Ed., (in press)
13. Moshkovich, A., Perfilyev, V., Verdyan, A., Popovitz-Biro, R., Tenne, R., Rapoport, L.: Sedimentation of IF-WS$_2$ aggregates and a reproducibility of the tribological data. Tribol. Intl. **40**, 117–124 (2007)
14. Moshkovich, A., Perfilyev, V., Lapsker, I., Fleischer, N., Tenne, R., Rapoport, L.: Friction of fullerene-like WS$_2$ nanoparticles: effect of agglomeration. Tribol. Lett. **24**, 225–228 (2006)
15. Allen, T.: Particle size measurement, 3rd edn, p. 215. Chapman and Hall Ltd, New York (1981)
16. Tevet, O., Von-Huth, P., Popovitz-Biro, R., Rosentsveig, R., Wagner, H. D., Tenne, R.: Friction mechanism of individual multilayered nanoparticles. Proc. Natl. Acad. Sci., (in press)
17. Scheffer, L., Rosentsveig, R., Margolin, A., Popovitz-Biro, R., Seifert, G., Cohen, S.R., Tenne, R.: Scanning tunneling microscopy study of WS$_2$ nanotubes. Phys. Chem. Chem. Phys. **4**, 2095–2098 (2002)
18. Lahouij, I., Dassenoy, F., de Knoop, L., Martin, J.M., Vacher, B.: In situ TEM Observation of the behavior of an individual fullerene-like MoS2 nanoparticle in a dynamic contact. Tribol. Lett. **42**, 133–140 (2011)

ABOUT THE AUTHOR

Prof. Reshef Tenne was born in 1944 on Kibbutz Usha to parents from Eastern Europe who had immigrated to the then Palestine in 1934. He was raised in the Kibbutz, finishing high school in 1962 and subsequently spent one year volunteering in the youth movement. Soon after his compulsory military service (1963–1966), he started his academic studies at the Hebrew University of Jerusalem. His M.Sc. thesis (1969–1971) focused on the photochemistry of Europium ions under the tutelage of the late Prof. Gabor Stein. His Ph.D. (1972–1976) thesis on the statistical mechanics of solutions was supervised by Prof. Arieh Ben-Naim and Prof. Shalom Baer. In 1972 he was married to Lea (Yonas) and their first child Dana was born in 1975. The next three years (1976–1979) were spent at the Advanced Studies Center of Battelle Institute (Geneva, Switzerland). In the first two years he was a postdoctoral fellow with Dr. Eric Bergmann (statistical mechanics of solutions). In the last year of his stay in Battelle, he became a member of the technical staff and started to develop new ideas in renewable energy research. His son Tal was born during this period (1979). In 1979 he joined the staff of the Weizmann Institute as a lecturer in the Department of Plastics Materials Research. His main interest was photoelectrochemical cells (PEC), first from the II–VI family of compounds and CdSe in particular. He developed (jointly with Dr. Gary Hodes) the photoelectrochemical etching surface-treatment which resulted in record solar-to-electrical efficiencies (close to 13%) for those cells. Later on (1984) he became interested in the family of compounds with layered structure and WSe_2 in particular. Applying similar surface treatment approaches, he found that he could obtain record efficiencies ($> 13\%$) for such cells. He became associate professor with tenure in 1985 and was promoted to full professor in 1995. His second son Ron was born in 1982. His first wife and the mother of his three children Lea